BUILDING YOUR OWN HOME

BUILDING YOUR OWN HOME

MURRAY ARMOR

Prism Press

PRISM PRESS,
2 South Street,
Bridport,
Dorset DT6 3NQ

and

PO Box 778,
San Leandro,
California 94577

Distributed in the USA by
Avery Publishing Group Inc
350 Thorens Avenue
Garden City Park
N.Y. 11040

First edition 1978.
Updated editions 1980, 1981, 1982.
Completely revised edition 1983.
Updated editions 1984, 1985, 1986, 1987.

ISBN 0 907061 97 4 Hardback
ISBN 0 907061 98 2 Paperback

Printed and bound in Great Britain by The Guernsey Press Co. Ltd.
Guernsey, Channel Islands.

Contents

Introduction

Building Your Own Home is both about self-builders and for self-builders. There are more of them than is generally realized, and they deserve more recognition than they receive. I hope this book will help obtain it for them, and that it will also prove a useful source of hard facts for those who are thinking of joining their ranks.

Building costs of one sort or another are quoted in nearly every chapter and great care has been taken to check them. Where they are actual costs of building a specific house or bungalow the year of construction is given, and where they are general figures they are current ruling costs. As far as possible I have quoted sources, so that readers who wish to up-date my figures will be able to do so.

The units of measure used are a mix of metric and imperial. This seems a muddle, but it follows the building scene today where drawings and regulations are in metric units, but imperial units are still used in general conversation. A self-builder will talk about using 4 inch nails to fix rafters at 600mm centres. I have followed suit.

All the case histories in the book are the real stories of real people, correct in every detail except that some names have been changed to respect the privacy of the persons concerned. I am grateful to them all for their generous help in making their plans, accounts, and other details available to me. Many others have also offered valuable help, and their assistance is gratefully acknowledged elsewhere.

Finally, it is fashionable in introductions to pay tribute to the understanding and tolerance of the author's wife while all was being written. In my case I would like to pay tribute to my wife's ruthless determination that I should finish this book and her insistence that I should keep it moving forward. This will strike a chord with many self-builders, for with both books and houses, the last lap is often the longest.

Introduction to the Ninth Edition

This ninth edition of Building Your Own Home is being written at a time when the Self-build movement is building more homes than ever before, and when Community Architecture is recognised as having an important contribution to make to the regeneration of run down urban areas. Both Self-build and Community Architecture provide ways in which the ordinary citizen can be involved in a new home, although surprisingly they have little else in common. Never the less, publicity for either approach helps the other by emphasising how important it is that opportunities are made for people to involve themselves in the bricks and mortar of their environment. Grants for Improvement Areas in the inner cities and opportunities to buy plots in the suburbs both meet the same need.

The new case histories in this edition include the adventures of families who were introduced to the idea of self-build by reading earlier editions of the book, and who wrote to the publishers offering to explain how they had made it all happen. I am very grateful to them and to all the others who have provided the facts and figures around which the book is written.

Murray Armor

Statistics on Self-Build housing completions in 1986

Hansard, 20 November 1986

314 *Written Answers*

NATIONAL FINANCE

Value Added Tax

Mr. Heddle asked the Chancellor of the Exchequer what was the number of value added tax refund claims under Customs and Excise Notice No. 719 processed for the fiscal year 1985-1986.

Mr. Brooke: The number of claims under Customs and Excise Notice No. 719 (Refund of VAT to do-it-yourself builders) processed in the fiscal year 1985-86 was 9,437.

Those who build for themselves without employing a builder are able to reclaim the VAT paid on the materials which they have purchased by using the provisions of the regulation to which the parliamentary question refers. They may only make one such claim, so that the total of claims indicates the number of homes completed where the VAT has been reclaimed in this way. As detailed in the reply above, the total in 1986 was 9437.

An estimated further 800 houses were completed by members of self-build groups who registered their groups for VAT as businesses, and who reclaimed their VAT on a monthly basis.

Other self-builders are small businessmen who reclaim their VAT through their own business accounts. These are people like smallholders, nurserymen, self-employed contractors, small farmers, etc, etc, and from the proportion of them to the number of VAT notice 719 reclaimers who are clients of companies who provide them all with services, it is certain that they account for at least another 1000 new homes.

The final total is thus estimated at 11,200 new homes completed by self-builders in 1986. This puts self-builders at the top of the league of house builders in the country for 1985. *House Builder* magazine in February 1986 gave estimates for the top three commercial house builders at 10,000, 9,600 and 8,000 respectively.

Build Your Own Home

Over 10,000 houses were built last year by D.I.Y. house builders, some on their own and some as members of self-build groups. Some literally build themselves, some do only part of the work, some provide only the management. The houses are better constructed than the average with savings of up to 40% on builder's prices. This book looks at the self-build scene, explains what is involved in building your own home, and describes exactly how it can be done in this age of regulations and restraint. It is full of options, schedules of requirements, programmes and financial planning. Somewhere there is mention of the need for hard work. Involvement in over two thousand self-build homes in twenty years has taught that this order of priorities is correct, and that if one is to build a home for oneself the project planning is everything, and compared with it the construction work is almost fun.

For some it is also something of a crusade, as a special sort of self-build also operates within the community architecture movement. This is enthusiastically promoted by those who fervently believe that one answer to many of societies problems lies in the individual being very closely involved with every aspect of his or her immediate environment, including being involved in their own housing from the planning stage to the actual construction work. In recent years this has led to a number of unusual self-build schemes in inner cities, most of them for those in housing need. One of these is described at length in a case history at page 105, and others are examined in a special chapter on inner city self-build later in the book.

These schemes rightly attract a great deal of media interest, although it is unlikely that in total they accounted for more than 100 new self-built homes in 1986 out of the grand total of 11,200. They have an importance out of all proportion to the numbers involved, and this sector of the movement is growing faster than any other. However, the philosophy and motivation behind much of it is so very special that it is described entirely separately in its own chapter, and what follows here deals with the 'conventional' self-builders who are the overwhelming majority.

For most house owners the purchase of their home is the largest transaction of their lives, and the decision to buy a house is a step only rivalled in importance by changing jobs or getting married. A publicity machine promoted by builders, estate agents and building societies helps them to their decision, and guides and assists them in the details. To reject all this professionalism is to flout the conventional, and before looking at how it is done it is important to understand what motivates self-builders.

Self-build offers great savings and this often leads to an assumption that self builders build for themselves because it is the only way they can own a home, get out of rented accommodation or escape living with in-laws. This is very wide of the mark. Any new property, self-built or not, is far more expensive than the cheapest existing housing in any area, and self-build is most popular in areas adjacent to industrial towns where Victorian terrace houses, with good restoration/capital appreciation possibilities, are available. If it is true at all it has to be qualified to read that self build offers some the only way in which they can get the home that *they want* within their means. What they want is invariably the best.

There is virtually no economy self-build housing, either in terms of size or specification. Most Housing Associations build larger properties to a higher specification than developers' estate houses. In their specifications they tend to take a large view with twice as much insulation in the roof, insulated floors, and pipework installed for central heating even if the appliances cannot be installed within the budget. This is done with resources which could be more easily used to buy something smaller and nastier. Not surprisingly, they do not see their situation in this light, for to sustain them in the effort which they are making, they have decided that their new home is the minimum that they are prepared to accept. Time after time one is told that 'this is the only way I can afford to build', when what is being built will make the self-builder the best housed individual in his family or among his work colleagues. Farmers are masters at this form of self-deception, persuading themselves that by building on a direct labour basis, they are just managing to afford to get a roof over their heads, when the roof is of a size and standard only normally seen in the stockbroker belt.

The real motivation in self-build is far deeper than costs per sq. ft. We were cave men for 20,000 years before Estate Agents arrived, and the urge to find one's own cave is deep rooted. In a world where opportunities for self-expression are diminishing, building one's own home both satisfies primaeval urges and exorcizes modern frustrations—with enough obstacles to be overcome to provide a sense of achievement. The rationale may be economic: the motivation is more basic. The new home is an expression of the individual himself, and will be the highest standard to which he can

The Self-Builder's Savings — A Comparison with Developers' Costs

Developers' Costs	*Self-Builder's Costs*
1. Land Cost.	1. Land Cost.
2. Interest on land cost over a long period.	2. Interest on land cost over a very short period.
3. Design and planning fees.	3. Design and planning fees.
4. Site labour costs.	4. Site labour costs, which may be as low as the developer's or could be at premium rates.
5. Labour overheads — cost of labour between profitable jobs, in periods of bad weather, holidays, Training Board levy, N.H.I. etc.	5. No labour overheads.
6. Materials at trade prices.	6. Materials probably at trade prices.
7. Up to 10% of materials damaged/wasted/stolen on site.	7. No site losses.
8. Office overheads.	8. No office overheads.
9. Staff costs and staff overheads.	9. No staff.
10. Expensive general contractor's insurances, N.H.B.C. guarantee fees, Trade Association levies.	10. Cheap simple site insurance.
11. Sales costs.	11. Nil.
12. Provision for bad debts.	12. Nil.
13. Interest on building finance assuming worst sales situation.	13. Interest on building finance kept to a minimum.
14. Corporation tax or other revenue involvement.	14. Nil.
15. Return required on capital.	15. Nil.

aspire. At the very least the standard will be several levels above the estate developer's lowest common denominator.

The level of individual involvement in self-build varies enormously. At one end of the scale we have the man who may contribute a thousand hours of manual labour as a member of a self-build group, at the other end the farmer or business man who contributes only management, employing self-employed workmen for all the building trades on a direct labour basis. The former contributes muscle and skill but has the backing of his association to deal with technical, legal and financial matters; the latter has to stand on his own feet. This involvement, in cash terms, means a saving of anything from 15% to 40% on the market value of the property. This saving is not simply the builder's profit, or the value of the self-builder's own labour, although both contribute to it. It really comes from a different cost breakdown for the building, shown below, and can be around £10,000 on a £40,000 house. This figure is so high that it can lead to unjustified criticism of the building industry. Of course builders make profits, but they also make losses, and the rate of bankruptcies is higher among building companies than in any other sector of industry.

During the 1890's and between the wars a number of groups of people in Britain shared a common social philosophy and built villages in which to live together to put their theory into practice. Today they would be described as communes. The best known are extensively documented in studies of the housing scene, and there are some who hope to see Self-build Associations as their successors. Those with this approach today are described in a later chapter, but the great majority of self-build groups are concerned solely with building the houses that they want within their budgets. Individual self-builders tend to be strong characters and often have strong political views of either the right or the left, but they certainly do not see that in building for themselves they are doing anything socially significant. They may be wrong, but they would be surprised to be told so. Those trying to live out new philosophies of living are involved with alternative housing, squatting or living in rural communes. The self-builder is trying to do his thing within the system, not trying to escape from it. His boast is of what his house is worth, and he usually builds a marketable house of conventional appearance and layout.

There is a clear distinction between those who build for themselves on their own, and those who do so as members of groups. The latter receive assistance from a variety of bodies and enjoy a special legal status, access to loan funds, assistance in finding land and other advantages. Individual self builders on the other hand are on their own, and the only authority which recognises them is the Customs and Excise, with special VAT regulations for those who are described as 'do it yourself housebuilders'. In spite of this the individuals outnumber the group members. There are many types — taciturn hill farmers building at the end of a stony track where no builder will quote for the job, businessmen building to indulge a feeling that they would like to organise something for their own benefit as a relaxation from doing the same for a company, and young couples determined to own a new house paid for out of a weekly wage on which they cannot afford anything the developer can offer. Some have no option but to build for themselves, such as our hill farmer and those who farm crofts or live on islands, but there are relatively few of them, and their stories usually merit a book of their own.

It is convenient to divide self builders into three separate groups, and it is important to recognize the differences between them as the way in which they can finance a new home are quite different. They are:

Those who have a plot of land with an unconditional planning consent, or who already have a plot on which they can get this sort of planning consent. An unconditional consent is one which does not define who can or cannot live in the dwelling, and so the property to be built is an unencumbered asset, which can be freely bought and sold. As such it will qualify for an ordinary mortgage, and because the self builder already owns the land he can use this as security for building finance. More of this later: at this stage the essential feature of this situation is that having already got the land, the self builder can hope to borrow all or most of the money to build on it.

Those who have a plot of land, with conditional planning consent. The conditions will link occupancy of the property to the land, requiring that the person living in it should either work on the land, or should be a pensioner who used to work on the land, or the widow of someone who worked on the land. This category of self builders covers farmers, smallholders, those who run caravan parks, dog kennels etc. etc — anyone whose job requires that they live where they work. The dwelling cannot be sold separately from the enterprise with which it is linked, and so it does not qualify for an ordinary mortgage and will have to be financed as part of the enterprise itself.

Those who do not have any land on which to build. These self builders have first to find a building plot, and then they have to find the funds to buy it. Once they have bought it they can use the title to the land as security to borrow their building finance, obtaining either a Bank or a Building Society mortgage. Unfortunately, they will find it very difficult to borrow money to buy the plot in the first instance, although this can be arranged if they have collateral in an existing dwelling i.e., if their home is worth a good deal more than the existing mortgage on it.

By definition, the self-builder is someone who contributes at least the management involved in building his own house, and sometimes a great deal of labour. An alternative to being a self-builder is to employ an architect to deal with all aspects of design and construction. This will include the drawings, dealing with all the technical and legal formalities, arranging for a contract to be signed, and supervising the construction process. The only initiative required from the client is to find the right architect to act for him, and following this decision he can take as much or as little interest in the project as he wishes, as the responsibility for everything rests with the architect. However invariably there is a continuous dialogue between architect and client throughout the duration of their contract.

The architect will be paid a fee of about 7% of the finished cost of the building, plus expenses and other charges. As for building costs, while the architect is concerned to achieve proper professional standards at the lowest reasonable cost, he operates at a level in the housing market which results in the finished cost per square foot being anything from 50 to 100% higher than those of the self-builder.

The fact that the building has been architect-supervised, with progress certificates to prove it, is accepted by the Building Societies as evidence of quality for mortgage purposes. If there are culpable defects or problems then the architect, his insurers, and ultimately his professional association will accept the responsibility. If this is the service you require then you are not a self-builder, but a prospective architect's client. The important thing to remember in making a success of the architect/client relationship is to ensure that the architect knows exactly what the total budget is. Unless this is clearly defined at an early stage it is possible to get deeply involved with designs for a building which is far beyond your pocket, with ultimate disappointment and recriminations. The frank disclosure of financial realities should be reciprocal—the client should state clearly what he is prepared to spend and the architect should be able to give a budget figure per square foot for a dwelling on a particular site.

The alternative to using an architect, but still avoiding the management of the project, is for the site owner to employ a 'design and build' organisation specialising in single residential units. These vary from local builders who can offer a range of designs from a book of plans, or who have draughtsmen or staff architects to draw up plans for clients, through to national companies who offer a highly specialised service. The advantage of dealing with such companies is that they are commercial and everything will have a price tag. This sets the tone of discussion, and the prospective purchaser will find that he is less inhibited in asking questions, and will probably be taken to see show houses or other recently completed buildings. Provided that he realises that he is dealing with a professional marketing organisation this is all to the client's good.

The houses offered by these companies are well designed and with their experience they are able to advise on the designs likely to be approved by the planning authorities. Where they differ is in the system of construction which they offer, and the consequent costs. Most companies in this market specialise in timber framed buildings which are quickly erected by specialist erection gangs. The shell of the building, roofed and waterproof, is given an outer skin of brickwork or masonry built in a conventional way. The finished appearance is usually traditional, and the timber frame construction enables a high level of thermal insulation to be 'built in'.

In the last year or so there has been a great debate about the merits and demerits of timber frame construction, with claims that timber frame houses are liable to condensation inside the walls and that they will be prone to problems as they get older. The timber frame manufacturers have responded by a very vigorous publicity campaign to refute these allegations, pointing out the very low level of warranty claims for timber frame buildings over many years. If all this helps to direct public awareness to the need for high housing standards this can only be a good thing.

By late 1984 timber frame construction had dropped to 12% of the UK housing market and has only a limited appeal to self builders. As will be seen from the case histories in the second half of the book, the lowest construction costs are invariably achieved by those who build traditionally. Wooden houses definitely tend to cost more when built by someone building for themselves. The same is generally true of concrete panel homes, which also have a public relations problem.

An alternative to timber frame or concrete panel homes are packages of designs and materials for traditional construction. The firms offering these are principally concerned with providing this service to smaller builders and Local Authorities, but are well used to supplying single sites. They will often design houses or bungalows to a client's requirements, and will find local builders, used to their service, to handle the construction work. They use traditional materials—brick, stone and insulating block—and have links with the manufacturers of these materials. Traditional materials guarantee a high resale value. The success of these companies enables them to work on smaller margins than the system building companies, and they claim significant cost advantages by using this type of service. Design and Materials Limited of Worksop are particularly active in this field, and publish a cost analysis every month giving the total cost of a recently completed house or bungalow for which they have been responsible. They have their own qualified staff architects, and both find builders for clients and also provide their service to those who are building themselves.

Those who do not entrust their requirements to an architect or building firm are self-builders and the rest of this book should help them in planning the whole project. Those who are using professionals should find it of value in arranging to get the right service at the right price.

The Planning Scene

A person intending to build a house must understand the various approvals necessary to build a dwelling. Dealing with these approvals, without which nothing can be legally built and which are essential for outside finance, is a job for a professional. This is normally an architect, either in private practice or working for a firm providing a package deal service, but can also be a design firm which is not a registered architect's practice, or sometimes an individual with relevant experience who submits applications as a part time job. Getting the application right is important and an understanding of what is involved will help in choosing the right man to handle the matter.

Planning Consent

The Planning Acts are concerned with whether or not a dwelling can be built at all in a particular locality, and, if it can, with its appearance and the way in which it will relate to its surroundings. This control is exercised by the local authority, to whom an application has to be made to erect any new dwelling. Members of the public are entitled to appeal to the Minister for the Environment against any decision of the local authority in a planning matter. In theory all planning applications are considered by a committee of councillors who are advised by the council's professional planning officers, but in practice they simply rubber stamp the recommendations of the planners for run of the mill applications. The planners make their recommendations in accordance with set criteria after going through set procedures. The way to get a planning consent quickly and easily is to ensure that it meets all the established criteria for an approval. This is often forgotten. A planning officer does have discretion to make recommendations which are at variance with planning policy, but this is unusual. With the planning situation on a potential building site an essential ingredient of every self-build project, it is important to avoid applications which may be contentious.

It is always the land which acquires planning consent, not the person who applies for it. When land with planning consent is sold, the consent is available to the new owner. One does not even have to own a piece of land to apply for planning consent on it, and it is quite usual for a prospective purchaser to make an application on a site which he is buying before he obtains the title to it, although he is legally obliged to tell the owner about the application. There are various types of consent, and it is important to appreciate the differences between them, which are quite specific.

Outline Consent is authority for a dwelling of some sort to be built, and is normally granted subject to a large number of conditions or 'reserved matters'. These will require that all details of design, materials, access etc. should be the subject of a subsequent application. An application for outline consent can be made before the expense of drawing plans is incurred.

Approval of Reserved Matters is given on applications to build a specific house on a site which already has outline consent.

Full Planning Permission is granted where applications cover all the matters regulated by the planning authority in one application. There can be reserved matters in a full consent, but these are always minor issues such as the type of brick to be used.

Conditional Consent is normally met as a planning consent for a farm cottage which is conditional on the dwelling being occupied by a person employed in agriculture. The actual wording used is often critical and might read: 'Occupation shall be limited to a person solely or mainly employed in agriculture, or last so employed, or the widow of such a person."

Conditional consents have a fascination for prospective self-builders, who often have ideas of establishing a connection with agriculture to take advantage of them. This is not recommended to any but genuine sons of the soil because long term finance for a building with such a consent is usually only obtainable from agricultural credit sources. Also the fascination may lead to building a house and being in contravention of the consent, and the track record of local authorities in winning confrontations in such situations is excellent.

A Planning Application has to be acted on by the planning authorities within a set period and once a decision has been made it is on record. However, an application can be withdrawn by the applicant before a decision is made which in some circumstances is useful.

A Refusal of a planning application is a matter of record, but does not preclude a further application being made to do the same thing in a different way. However, a refusal of an application for approval of reserved matters can mean that any re-submission must be as a full planning application, so it is usual to withdraw such an application before a decision is made if bad vibrations are detected in the planning office. More of this anon.

Conservation Areas. These are areas of particular architectural importance where all proposals are subject to special scrutiny and where any new building is expected to blend in perfectly with its surroundings. Conservation areas are usually found in town centres, in particularly attractive rural villages, and in beauty spots. Building in a conservation area can cost a great deal over the odds, but the resulting property is likely to be worth correspondingly more. Building plots in conservation areas rarely come on the market, and the prices they command are high.

Green Belt, White Land, Residential Zoning etc. These are phrases which are often used in different ways and are usually not very specific. They relate to existing development plans for towns or villages and are used as a basis for dealing with planning applications. 'Green belt' refers to land where no development is permitted and 'White Land' is land not yet zoned for residential development. This seems straightforward, but only small areas of the country have development plans, and these are constantly being redrawn. Furthermore they are regarded by different authorities as anything from a vague statement of principle to Holy Writ. Terms such as these are really only specific to a particular development plan, and the standing of that plan can vary. None of this is very helpful.

Appeals. Self-build is difficult enough without looking for complications. Avoid applications which are unlikely to succeed. If you do fail, avoid appealing. If you wish to disregard this, take professional advice.

When one looks at the different types of consent it seems logical to apply first for outline consent to establish whether a house or bungalow can be built at all, and then, only when this is granted, to move forward to preparing detailed plans and submitting an application for approval of the reserved matters. This is frequently advocated by estate agents, solicitors, and accountants as it facilitates planning a client's affairs without committing him to detail. However, this approach can be disadvantageous.

Architects often prefer to make a full planning application. This saves time, and can demonstrate that an application is bona fide, and not simply made to establish an enhanced value for a piece of land. Architects also suspect that planning officers may impose expensive conditions in an outline consent, while they might have accepted a cheaper alternative had it been presented to them as part of a full application. For instance, an outline consent for an infill site in a rural village might be qualified: 'The building shall be of two storeys, shall conform to the building line of adjacent properties, and shall match them in style and character'. It may have been the applicant's intention to build a bungalow hidden away at the back of the site, and this possibility may have been over-looked by the planning officer. Any subsequent application for the bungalow will be judged on its merits, but the planning officer is unlikely to see anything which did not conform to his original qualifications as an improvement.

Whether or not to make an outline or a full application can be a difficult decision and can depend on the council involved. Some rural councils have so few applications that the planning committee is able to see the drawings for most new dwellings, and a well presented set of drawings, with a perspective sketch, must create a favourable impression at first sight. In areas where the drawing is likely to be seen only by the planning officer, the presentation is relatively less important. Be guided by your architect, or whoever is to submit your plans.

The public have free access to planning officers, who invariably have set times when they are available to discuss the relevance of planning policy to specific sites in their areas. Their treatment of visitors is more friendly and open than is usual in local government. However, in dealing with them recognise two things. Firstly, they cannot discuss planning policy, but must restrict themselves to advising how planning policy will affect your proposals, without committing themselves or the council in any way. A planning officer will advise that an application 'could normally be expected to be approved' or that he 'would not anticipate being able to recommend approval'. He is not allowed to be more specific, and it is pointless, and rude, to try to tie him down. What he will do is explain planning policy in such a way that there are plenty of lines to read between. What he will not do is debate the policy itself, or give any sort of promise regarding an application. The exception to this is when considering minor reserved matters in a full consent, such as a requirement that a type of brick should be approved. Approval of these matters is delegated to the planning officer by the council and here he will always be specific and will often negotiate and compromise.

The second point to consider before entering the planning office is that if you ask for advice, you will get it. Planners have an interesting job, and almost without exception are interested in it. They are anxious to influence development by advice as well as by control, and there is usually a notice at the reception desk saying that they are always ready to advise on proposed developments. This means what it says. Unfortunately this advice is usually a counsel of perfection and it may not suit you to take it. For instance, suppose you are buying an attractive site with outline consent. As a first stage in establishing a design it would seem sensible to call at the planning office to ask for free advice. You will be well received, and will be impressed by the trouble taken to explain how 'the site needs a house of sensitive and imaginative design to do justice to its key position' . You may be shown drawings, a sketch may be drawn for you, and it is not unknown for the planning officer to drive to the site with you. This is splendid, until you realise that the ideal house being described suits neither your life style nor your pocket. You will wonder what the reaction is going to be to your application for the quite different house that you do want, and which is all you can afford. The simple answer is that your application will be dealt with on its merits, and that the planning officer has an obligation to approve what is acceptable, and not what he thinks best. However, presumably he will be disappointed and this disappointment must colour his thinking.

For office use only

Ref No	D	C
S		
Land Use Classification		D

the accompanying notes before completing any part of this form

ON FOR PERMISSION TO DEVELOP LAND ETC

ountry Planning Act 1971 — Date received

s of an application form and supporting plans (or four in the case of land in a parish) shall be submitted

o be completed by or on behalf of all applicants as applicable to the particular development.

(in block capitals) — **Agent** (if any) to whom correspondence should be sent (in block capitals)

Name ....
Address ....
....
Tel. No. ....

rs ot proposal for which permission or approval is sought

address or location of the land to which application relates and site area (if known)

the land owned by the Applicant on 12th September, 1974 — YES/NO

articulars of proposed development ng the purpose(s) for which the land buildings are to be used.

hether applicant owns or controls any ng land and if so, give its location

hether the proposal involves - *State Yes or No*

ew building(s)

teration or extension

hange of use

If residential development, state number of dwelling units proposed and type if known, e g houses, bungalows, flats and number of habitable rooms bed spaces *(see note 16)*

onstruction of a new ccess to a highway — vehicular / pedestrian

teration of an isting access to highway — vehicular / pedestrian

rs of Application (see note 3)

e whether this application or - — *State Yes or No*

ine planning permission .... →

planning permission .... → — If yes, delete any of the following which are not reserved for subsequent approval: 1 Siting; 2 Design; 3 External appearance; 4 Means of access

roval of reserved matters owing the grant of outline ission .... → — If yes, state the date and number of outline permission. Date .... Number ....

ewal of a temporary permission or permission for ntion of building or continuance of use without comng with a condition sub to which planning ission has been granted .... → — If yes, state the date and number of previous permission and identify the particular condition *(see note 3d)*. Date .... Number .... The condition

4. Particulars of Present and Previous Use of Buildings or land

State

(i) Present use of buildings/land (i)
(ii) If vacant, the last previous use (ii)

5. Additional Information

(a) Is the application for Industrial, Office, warehousing, storage, or shopping purposes? *(see note 5)* — **State Yes or No** — If yes, complete Part 2 of this form

(b) Does the proposed development involve the felling of any trees? — **State Yes or No** — If yes, indicate position on plan

(c) (i) How will surface water be disposed of? (i)
(ii) How will foul sewage be dealt with? (ii)

(d) State details of fuel burning apparatus to be installed.

6. Plans

List of drawings and plans submitted with the application

NOTE: ***The proposed means of enclosure, the materials and colour of the walls and roof, landscaping details should be clearly shown on the submitted plans, unless the application is in outline only***

* I/We hereby apply for

*(a) planning permission to carry out the development described in this application and the accompanying plans, and in accordance therewith.

OR *(b) planning permission to retain buildings or works already constructed or carried out, or a use of land already instituted as described on this application and the accompanying plans.

OR *(c) approval of details of such matters as were reserved in the outline permission specified herein and are described in this application and the accompanying plans.

*** Delete whichever is not applicable** — Signed ....

On behalf of .... (insert Applicant s name if signed by an Agent)

NOTE: ***An appropriate certificate must accompany this application unless you are seeking approval to reserved matters — see Note 10. The following certificate will be appropriate if you are the "owner" of all the land***

CERTIFICATE UNDER SECTION 27 OF THE TOWN AND COUNTRY PLANNING ACT 1971

Certificate A

I hereby certify that -

(a) "owner" means a person having a freehold or leasehold interest the unexpired term of which was not less than 7 years

1. No person other than the applicant was an owner *(a)* of any part of the land to which the application relates at the beginning of the period of 20 days before the date of the accompanying application.

*2 None of the land to which the application relates constitutes or forms part of an agricultural holding. or

*3 *I have / The Applicant has — given the requisite notice to every person other than *myself / himself who, 20 days before the date of the application, was a tenant of any agricultural holding any part of which was comprised in the land to which the application relates, viz -

(b) If you are the sole agricultural tenant enter "NONE"

Name of Tenant *(b)*	Address	Date of Service of notice

Signed ....
On behalf of ....
Date ....

****Delete either 2 or 3 (NOT both) whichever is not appropriate***

A typical planning application form. All authorities have slightly different forms — goodness knows why — and it is most important that they are filled in correctly. The standard Certificate A shown at the end of this form is simple to complete, but if you do not own the land, or if it is an agricultural holding, then Certificate A does not apply and you will get involved with a lot more paperwork. Normally the architect or the firm that you employ to make the application will deal with all of this.

From this is seen that a preliminary discussion with a Planning Officer is not to be taken lightly. As a general rule, the best person to deal with the planning officer is the professional who you employ to submit your application.

A fee is payable when submitting a planning application, and has to be paid when the planning forms are deposited at the council offices. The standard fee is payable whether it refers to an outline application, a full application, or an application for approval of reserved matters. There is no refund if the application is refused!

Typical Negotiations on a Planning Application — What to Expect

The following is an extract from an actual letter received as a result of a straightforward application made to a West Country local authority.

> "I have a number of observations to make regarding this application, and you may wish to comment on them before I make my recommendation to the planning committee.
>
> 1. While I appreciate that the dwelling has been sited to obtain a veiw from the ground floor windows over the trees to the south east, it is now well behind the building line and this does little for the street scene. I suggest that moving the building forward by 4 metres would effect a satisfactory compromise between user and environmental considerations.
>
> 2. The proposal to fell the existing horse chestnut tree in the proposed driveway, and to replace it with a new tree 2 metres to the south, is only acceptable if the proposals specifically relate to a well established indigenous tree of an approved species.
>
> 3. The proposed visibility splay to the access is not in accordance with the requirements of the county highways department for an access into a road with the width of less than 6 metres. Presumably details of this requirement are available to you, but if not they can be obtained from County Hall.
>
> 4. The proposed roof pitch of 35° is not acceptable and I am not prepared to recommend approval of any roof in this village with a pitch of less than 40°, 45° would be preferable. This also applies to garages, even when situated well behind the building line.
>
> 5. While the proposed fenestration is in character with adjacent buildings the pair of french windows is completely out of character. Cannot they be moved into the rear elevation? If access to the side garden from the lounge is considered essential, a single glazed door would be preferable.
>
> 6. Are the soldier course of bricks above windows as shown on the drawings to be finished as segmental arches as in the chapel on the opposite side of the street? This is not clear from your drawings, but would be welcome.
>
> 7. To the best of my knowledge the bricks proposed have not been manufactured since 1976. Is your client proposing to use second hand bricks of this type? Alternatively a full specification and samples of the bricks proposed will be required.
>
> As the proposals stand I do not feel that I can recommend the committee to approve your application. However, if you can be available to discuss this I am sure that we can achieve a satisfactory compromise, and look forward to hearing from you with a view to making an appointment.
>
> Your faithfully,
> Planning Officer"

Meeting reaction of this sort from a local authority within an established budget, confident of winning points as well as having to concede others, is a job for a professional.

Building Regulation Approval

Approval under the Building Regulations is concerned with the design and siting of a dwelling from every angle except that of its appearance and whether it should be built at all, which are planning issues. A mass of health and public safety legislation is now administered by the local authorities through their building inspectors, and for many years it has been a requirement that drawings of proposed buildings should be submitted to local authorities for examination and confirmation that they conform to the building regulations. This confirmation is known as Building Regulation Approval.

In 1985 new Building Regulations were introduced which gave Architects with appropriate professional indemnity insurances, and other bodies such as the NHBC, the authority to inspect and certify the construction of new buildings as an alternative to the Local Authorities. The various bodies concerned have not yet published their arrangements for this, and it is likely to be some years before self-builders have any easily available alternative to the

Local Authority and Building Inspector route to Building Regulation approval. What follows is concerned with this Local Authority approval of plans, and the Building Inspectors approval of work on site.

Fees are payable with building regulation applications and it is usual for an application for building regulation approval to be made at the same time as the planning application. Approval of a Building Regulation application is a matter of fact — whether the drawings of a building conform to the regulations or not. Virtually every detail of the building, of its siting, and of its services are subject to the regulations and it is not enough to simply note on the drawing that everything will conform to the rules. All the salient points must be detailed individually.

Having checked these the authorities will invariably write asking for further details, and may suggest a meeting. Finally, they will issue an approval form, which states that the plans inspected conform to the regulations, together with a set of postcards to be sent off to the council at various stages as construction progresses. On receipt of these cards, and at any other time he thinks fit, the Building Inspector will call at the site to

BUILDING CONTROL
PLAN No.
RECEIVED ....
O.O.T. ....

PUBLIC HEALTH ACTS 1936 and 1961
THE BUILDING REGULATIONS 1972

Notice of intention to erect, extend or alter a building, execute works or install fittings or make a material change of use of an existing building.

THE BUILDING REGULATIONS, SECOND SCHEDULE, RULES B, C, D, F, G

To – ....

I/We hereby give notice of intention to carry out the work set out herein

Signed ....

Date ....

Name and address of person or persons on whose behalf the work is to be carried out (IN BLOCK LETTERS PLEASE) ....

(Telephone No ....)

If signed by an Agent: –

Name and Address of Agent (IN BLOCK LETTERS PLEASE) ....

(Telephone No ....)

Name and Address of Owner of the property ....

Has the owner agreed to the proposed work? ....

1 Address or Location of proposed work ....

2 Description of proposed work ....

3 (a) Purpose for which the building/extension will be used ....

(b) If existing building, state present use ....

4 Means of water supply ....

5 Foul drainage will go to ....

S.W. drainage will go to ....

NOTES

(a) This Notice should be completed and submitted together with where applicable Plans in triplicate (one copy on linen) in accordance with the provisions of the Second Schedule of the Building Regulations.

(b) Additional information may be requested pursuant to Rule E of the Second Schedule to the Building Regulations.

A typical Building Regulations application form. This is a lot simpler than the planning form, as all the detailed information required by the Local Authority has to be noted on the plans themselves. This design of form pre-dates the revised 1985 arrangements, but is still in general use.

inspect the work. His concern will be that all construction complies with Building Regulations, and not simply to enforce the details shown on the approved drawings.

It is evident that Building Regulation applications, like planning applications, are best submitted by professionals. Once the consents have been granted there is no further contact with the planners, while liaison with the Building Inspector continues until the dwelling is occupied. At this stage it must be emphasised that for the self-builder his Building Inspector has got to be a friend. He is responsible for ensuring that construction is as required by the book, and he has no authority to deviate from it. He knows the book backwards, which gives him an advantage over anybody who has not made a special study of the subject. He spends his life dealing with those trying to ignore or bend the rules, and dealing with problem buildings. In spite of all this it is likely that he will have remained a sympathetic, friendly and helpful character, and can become guide, mentor and friend to the self-builder.

Occasionally self-build groups engage in feuds with their Building Inspector. This is not recommended; battles with Building Inspectors are battles lost before they are started. A Building Inspector whose inspections are welcomed and whose advice is sought by self-builders can be a tremendous help. Officially they will advise on any detail of the construction, and unofficially they will often recommend sub-contractors, suggest suppliers, and generally make their unrivalled experience available. No one else in the building industry sees all the work on all the sites in an area.

Incidentally, many building inspectors are self-builders. They have an enormous sympathy for the movement, although this may not be appreciated by those who get told to deepen their foundations by another six inches after

spending the whole day squaring off the bottoms by hand. I remember a self-builder in Dorset telling of his own experiences. He was building on wet ground near a lake, and the Building Inspector properly insisted on being present when the raft foundation was poured. The self-builder had only a Saturday and Sunday free for this work, and after making this point strongly to the council, the inspector was instructed to supervise the job over the weekend. At 8.00 a.m. on the Saturday he arrived at the same time as the first concrete truck, ready to watch the self-builder, helped by his wife, lay the concrete. By 10.00 a.m. he had been home for his old clothes and was in the foundations working with them. This level of assistance cannot be taken for granted, but the attitude is not unusual.

Negotiations on a Building Regulations Application — What to Expect

The letter below is a typical communication from a local authority following a Building Regulation application. It is in no way unusual, and irrespective of the amount of detail shown on the drawings submitted, queries of this nature can be expected from most local authorities.

> "I refer to the plans referenced above which have been deposited for consideration under the Building Regulations. Your attention is drawn to the following items regarding which I require clarification to enable further consideration to be given to this matter.
>
> 1. Further details required of the porch roof construction. Denote siting and method of supporting cold water storage cistern, with calculations to confirm that roof truss design is adequate. Detail wind bracing. Denote waste sizes to sanitary appliances.
> 2. Confirm arrangements made for cavity to be closed at cills and reveals.
> 3. Confirm flashings as shown to roof to provide 150 mm upstand.
> 4. Buttress walls to be taken down to similar foundations as external walls. Indicate strutting proposed at mid span to first floor joists over lounge area. The first floor gable return appears to be undersize, and if not in accordance with deemed to satisfy provisions, calculations will have to be provided. Detailed proposals for lateral restraint to gables at first floor ceiling and roof level. Detail size of trimmer below bulk head studding over stairwell.
> 5. Detail section of stairs, including head room to stairs measured vertically above pitch line.
> 6. Confirm that proposals for balustrades conform with amendments III Building Regs.
> 7. Confirm relationship between ventilation openings and floor area to bedrooms 1 and 4.
> 8. Full details of the prefabricated septic tank are required, and we will require a standard percolation test to be carried out in accordance with section 3/5/6.2 of Code of Practice 309 1972.
>
> I regret that unless I receive an amended application taking all the above matters into account I will have no option but to reject the plans in fourteen days from the date of this letter."

It will be appreciated that dealing with this sort of query, within the Building Inspector's deadline, which he cannot alter, requires professional advice in almost every case. An architect would consider all the queries raised above to be perfectly straight forward and would expect to deal with them as a matter of routine. Incidentally, the seemingly haphazard order of the matters raised in the letter above is the order in which they are dealt with in the regulations themselves.

Self-Build Projects

Owner built home in Wales.

Owner built home in Sussex.

Self-build Housing Association homes in East Anglia, built in the regional style.

Finance

Project Finance

Money is what self-build projects are all about—the opportunity to use one's own money so effectively that one gets a bigger or better house than it would buy on the open market. The management of this money is as important as the management of the building work, and far more hazardous. The Building Inspector and others will do all they can to prevent construction problems, but no-one is going to jump up and cry a halt if you are heading for a financial disaster. In this chapter we shall look at the financial plans that have to be made, and the safeguards that must be built into them. Of course, no self-build venture can ever be entirely safe; those building for themselves are moving into the capitalist jungle of the builder and developer, where fortunes are lost as well as made. What they can do is try to plan so that the odds are in their favour.

It is with finance that Housing Associations have most to offer. They have access to loan funds, to commercial credit, and have the backing of the Housing Association establishment. Their rules ensure that their book-keeping and financial planning is always up to date and members of associations have the strength of the group behind them. If a member wishes to leave he will find that he can do so without facing financial disaster, and he will be replaced without undue difficulty. However, although he has these advantages, the individual member of an association must still plan his own money affairs in relation to his obligation to the group, and he will contribute more to the group discussions if he understands about their finances. This is best examined by looking at the more complex problems of the individual self-builder.

When a project is to be financed by an individual, answers have to be found to the following questions.

1. Where is the money coming from to pay for the house when it is finished?
2. Where is the money coming from to purchase the site before a start is made?
3. Where is the money coming from to pay for the materials and labour as work progresses?
4. How is this managed so as to pay as little interest as possible?
5. How can insurance cushion the major risks?
6. How will this all be done without affecting one's normal pattern of living in a way that is unacceptable?

The fortunate minority already have the money available, in the bank. Even here financial planning is necessary to ensure that it is used to the best advantage, and that adequate provision is made to complete the building should a death or other circumstances change the situation. However these affairs are arranged, it is certainly desirable to make specific provision for the new building, with the funds invested where they can earn maximum interest, and where they can be withdrawn as required. There are many formulae for this; one is to place the whole sum on special deposit at 28 days notice through a bank, earning far more than ordinary bank deposit rates, and to use this as security to operate an overdraft on a current building account until heavy expenditure has to be incurred.

For many, particularly farmers and other self-employed people, a new home may be financed out of profits of a good year. Here make sure that the Inland Revenue do not present problems at a later stage, and again make specific provision for the building, checking with one's accountant that the money really will be available when it is required. This is even more necessary when a dwelling is to be built as a capital investment, especially on a farm where it can qualify for Capital Allowances. Again, think about the way in which it is to be built and if it depends on the management or labour of one individual, appropriate insurance cover or other provision should be made in case that person cannot finish the job. The same applies to planning for a house to be built from an inheritance before probate is settled.

A difficult situation can arise where a house is to be built using money provided as a gift. It is not unusual for a self builder to be offered financial help by a relative on a vague basis, and it may be difficult to insist that this welcome aid must be on the basis of a firm arrangement. Gifts of sums of money attract Capital Transfer Tax, and if this is not paid by the donor it can be charged to the recipient. There are exemptions and loop-holes, particularly if the recipient is getting married, or if the gift can be phased over a number of years, but to take advantage of this the advice of an accountant must be sought and acted upon. However obvious this may seem, these facts may prove difficult to explain to a great-aunt who announces that she will help you out if you run out of money before you finish. Still, if this is your problem, there will be many who will envy you.

Mortgage Finance

For most people the source of finance for the finished building is a mortgage. An understanding of what sort of mortgage one wants, what mortgage one can afford, and what mortgage one is able to get is a pre-requisite of all financial planning. Mortgages take various forms, but they all start off with someone, usually a building society manager or bank manager, deciding to offer you the help that you require. To do this he will wish to satisfy himself about your standing as a potential mortgagee.

He will want to be certain that you will make the mortgage repayments and will base his decision on his experience with people like you. A mortgagee is normally considered to be able to handle replayments of up to 2¼ years salary, although this can be flexible. The extent to which the bank or building society will gross up the earnings of a husband and wife for this purpose depends on many factors, and cannot be taken for granted. The sort of employment is important and tends to favour civil servants, teachers and others whose employment is a sinecure. Age is significant only as far as the obvious requirement that he or she should not retire before the end of the mortgage repayment period, and for the same reason they may not favour anyone in poor health. The ideal mortgage applicants are probably a town clerk married to a school mistress, while at the other end of the scale a professional gambler married to a club hostess would have problems. The way a building society considers the personal attributes of an applicant, as opposed to his status, is in considering his record as an investor. An applicant regularly saving a sum with the society roughly equal to the repayment on the required mortgage has proved his standing in the most convincing way.

Banks and Building Societies are in business to lend money and to receive repayments regularly, and regard having to foreclose on a defaulting mortgageee as their own failure. Their business is to spot bad risks, and not to risk their money with them. They must secure their loans and take a careful look at properties on which they intend to lend money. In this they are interested in two factors — the valuation and the 'percentage mortgage'. The valuation is the value of the property as assessed by the professional valuer appointed by the society. The figure which he quotes to the society is final and not negotiable in any way at all. The maximum mortgage which will be granted will be a percentage of this figure, and can vary from 95% on new properties with a guaranteed resale potential — typically three bedroom houses on a popular estate development — to 50% or less for old properties with a more restricted market. A building put up with a conditional planning consent is not considered an acceptable risk, as it cannot be resold on its own on the open market, but only with the farm or enterprise which goes with it.

If you are considering a bank mortgage you will presumably seek to arrange this with the bank that already looks after your account, and they will judge you on your track record. If you are considering a Building Society mortgage it is a very good idea to open a savings account with the Society and to discuss your mortgage with them as an established saver.

Peter Birch, chief executive of the Abbey National Building Society, cuts the first sod for the Endsleigh Self-Build Housing Association in April 1986. Inset left the first member to move into his new home moves in six months later. The houses were built at a 26.6% discount on their value.

Both banks and Building Societies are very helpful to registered self-build housing associations as seen from the photographs on page 17. They are equally well disposed to individuals building on their own, with the proviso that they will expect you to demonstrate that you are fully able to manage the job, and that you have every detail of what you want to do at your fingertips.

Knowing the size of the mortgage which you can hope to obtain is an essential first stage in your financial planning. Once you have established this you can move on to other things.

Site Purchase

Our next hurdle is the source of money to buy the site before work commences. Although a building society will advance 90% of the final cost of the land plus house, and although the site is perhaps 30% of the value of the finished home, they are unlikely to lend you the money for the site. The same is true of the bank. The exception is where you have some security to offer, such as a further charge on an existing home, and a reputation as someone who is going to succeed. The latter is as important as the former. Without it it is little use going to a bank and saying that your existing home is worth £50,000, your mortgage is only £25,000, and that you can offer a second charge on your home to cover a loan of £20,000 to buy a building site. The bank hates to get involved in failures, and it will recognise a risk that you will fail to build, fail to resell the land, and be unable to meet interest charges. Unless it knows from experience that you never fail at anything, it is unlikely to help.

Invariably the individual self builder has to find the money for the site himself, perhaps combining his savings, the proceeds of selling his car and buying a cheap van to help his building work, and perhaps with a loan from a member of his family who he can repay when he gets the mortgage. Some self-builders who already own a home sell it to finance the site and move into a caravan while they build. This has its own hazards, particularly as one gets out of the houseowner situation for a while. Every home owner knows the danger of this in a time of escalating house prices. If you decide to do this, and propose to live in a caravan, do not put it on the site until you have all your planning and byelaw consents, and are about to start work. Permanently occupied caravans require planning consent, and these are difficult to obtain. Certainly you do not wish to confuse the situation regarding the application for your house by also applying for a caravan, or to prejudice it by moving a caravan on to the site. When you have your consents then deal with the caravan issue separately.

Above all, avoid second mortgages or loan services of the sort which are advertised in local newspapers. Whatever the role of such finance in our society it is certainly not to provide finance for anything as fundamental as the family house.

Building Finance

Fortunately, the money to actually build is more easily obtained once you have a site, as the site can be the security for the loan. There are three ways of arranging this—progress payments from a building society, a bridging loan from a bank which is repaid by a building society mortgage when you complete the work, or a bank loan which is converted into a bank mortgage when you finish. A building society mortgage with progress payments has the disadvantage that the money is made available in stages; usually four installments, and until you get the first payment you may have no money to pay for the work to start the building. The usual stages are at damp proof course (or dpc), roofed, plastered out and completed. The payments will only be made after survey, submission of the surveyor's report, and subsequent paper work, so expect a fortnight's delay between completing a stage and obtaining a payment. As a result, even if your suppliers and any sub-contractors have agreed to wait to be paid until you get your stage payment, there will be an inconvenient lull in work on site at each stage while your creditors wait to see their money.

A bank bridging loan prior to a Building Society mortgage is more convenient, although more expensive. In essence, it is an arrangement for the bank to lend you all the money that you require to build, sometimes as a lump sum loan but preferably as an overdraft. As security the bank will want your personal guarantee and also a charge over the land, and whatever you build on it, until you finish when it will transfer its charge to your building society, and the building society will transfer its cash direct to the bank. The banks and building societies are used to working together in this way. The advantage of this is that the bank will be realistic about letting you have cash when you actually need it, although they will probably also want surveyor's certificates at intervals. The disadvantage is that the bank will want more interest, may insist on a loan instead of an overdraft which is more expensive still, and will sometimes demand a commitment fee as well. Tax relief can be claimed on this interest, but not normally until you make your annual tax return. Most self-builders use bank finance for building costs, and arranging this at an early stage is important. Before a bank manager will authorise a bridging loan he will want to see a letter from your building society confirming that a mortgage will be available in due course. It also encourages building society managers to hear that a bank seems likely to provide building finance.

A bank loan linked with a bank mortgage has obvious advantages, but bank mortgages are not always available and many self builders like to take a mortgage with a building society where they have an existing account.

Most self builders do a great deal of shopping around to sort out their building finance, and invariably you will have to set out to "sell" yourself and your ambitions to the banks or building societies that you approach. None of them advertise that they have loans available for self builders, and they will need to be convinced that your proposals are sound and realistic.

Insurances

Having made arrangements to borrow money you will suddenly be conscious that you have entered into a formidable obligation that can only be discharged by actually building the house. While this has always been your intention, the reality is sobering. You must ensure that you have adequate insurance to cover as many risks inherent in the building operation as possible; and also the risk of death or illness preventing you from finishing, leaving your widow or others with a half completed building and financial obligations. The latter is dealt with by taking out a short term life policy for a sum adequate to pay a registered builder to finish the building. The sum required depends on the circumstances but £6 per square foot of house would be adequate cover, and a healthy man under 40 would get 12 months cover on £10,000 for under £40. Health insurance depends on how you assess the risk, but it is also relatively cheap.

Building risk insurance is absolutely vital, as you must insure not only the building materials but also cover third party claims, and claims from sub-contractors and others who may visit the site. The law is not on your side, and you should realise that in general terms a child trespassing on your site who falls off your scaffold and is hurt has a valid claim for damages against you. Special insurance policies are available to deal with this, and a proposal form for one of them is at the end of this book.

Above: Self-Build Housing Association homes damaged by a freak gale while under construction. Uncompleted houses are very vulnerable to this kind of damage. Insurance had been properly arranged and the claim was met promptly and in full. The settlement enabled the Association to employ a contractor to do all the rebuilding.

Left: The same houses when completed.

Financial Planning

In this business of finance the group member is far better off than the individual. The group can negotiate both land and building finance in a way that the individual cannot, and helps with mortgage applications. The group also arranges for surveyor's reports or architect's progress certificates for individual members, and ensures that the actual building operations on site are adequately insured.

Both individuals and group members must ensure that their enthusiasm for their new house does not put unacceptable constraints on their normal finances. As a final exercise in financial planning it is important to look carefully at the effect of the commitments entered into on one's own life, and on the lives of one's family, and to make a conscious decision to accept these constraints. Involvement in a self-build project may mean that the cash flow situation will preclude holidays for a year or two, or that accustomed luxuries will have to be abandoned. The time to accept this is before you start.

The various financial arrangements described are set out on the project planning charts on pages 61-63. There are many different ways in which all of this can be arranged and the important thing is to see that it really is all *arranged* and not left to chance.

Project Cost

Having understood this frame-work — how much is a new house going to cost? This is a question to which there is no set answer. Design, foundation costs, the fittings and fixtures, transport charges and other factors all vary enormously, and inflation adds a further variable. Every prospective self-builder has to find out his own probable costs, and has to keep up-dating the figure to match inflation as his plans progress. To do this he will find some sources of information better than others, and some of no use at all.

Firstly disregard any analysis of the prices at which new homes are offered by speculative developers. Basically the developer sells a house for what he can get for it, and rarely sells any unit, or any estate of houses, at a price which is a set proportion of construction costs.

Other misleading sources of information are the quarterly statistics of housing costs published by the Government, local authority cost yardsticks for new Council housing, and the figures published by the Building Trade press and trade associations. All these figures are invaluable tools for the professional, but are of little use to the self-builder. Most houses are built on estates, and published figures relate to estates. They are built by large organisations enjoying the economies of scale, but with enormous overheads. None of this is relevant to your new home on your plot of land.

There are two sources of really useful information: personal contacts, and the cost figures put out by the package companies and the Housing Associations. Individual self-builders will often quote their building costs with pride. Information of this sort is invaluable. Bear in mind that these costs only reflect one job, and assess their reliability. Housing Associations have tight budgets which are continually up-dated, and rarely have inhibitions about divulging them. The problem is that the members are almost certainly too busy building to deal with casual enquiries. Try a visit to a site, or find the pub used by the association committee.

More readily available, and more reliable, are the construction costs quoted by those who offer a package service. The commercial companies offering this sort of service quote average costs, and frequently justify them by giving details of the actual costs of recently completed projects. Design and Materials Limited publish a cost analysis of a newly completed house each month, with plans, an outline specification, and a resumé of how it was built. These are almost essential reading for anyone interested in self-build.

Architects and builders will usually be hesitant about quoting prices, although local architects with experience of single housing units on individual plots should be able to quote guideline costs per square foot. Builders will tend to quote whatever costs they think the questioner would enjoy hearing, hoping it might lead to a contract. This is not because they are devious, merely that they have to survive in a harsh commercial environment, and without a site, drawings, consents and money in your pocket you are really making conversation rather than asking a question.

Whatever figures you obtain will be related to one way of building. The approximate costs involved in building in different ways have been analysed by D & M who say:

> 'February 1986 national average costs for bungalows and houses to D & M standard designs, built on single sites by individual clients, on straightforward foundations, including fittings and fixtures appropriate to the size of the property, central heating, double glazing, connection to drains or septic tank, garage, short length of drive, no landscaping.
>
> When built by a well established N.H.B.C. builder, working from his offices, following a formal invitation to tender, formal contract.
> *—£30 to £36 per square foot (January 1986 contracts).*
>
> When built by a reputable small builder, N.H.B.C. registered, working from his home, usually himself a tradesman, following an informal approach, contract established by exchange of letters.
> *—£26 to £32 per square foot (January 1986 contracts).*
>
> When built by a competent private individual on a direct labour basis, using sub-contractors, without providing any labour himself.
> *—£22 to £25 (January 1986 completions. Occasionally much less).*
>
> Self-build housing associations, over 80% of labour provided by the association members.
> *—sometimes below £20 sq. ft.*
>
> The number of square feet of floor area in a property is the area enclosed by the internal faces of the external walls, (which is very different from the plinth area).
>
> The average figures quoted remain remarkably consistent irrespective of the size of building or whether it is a house or a bungalow with the more expensive fittings in larger properties being balanced by savings consequent on the economies of scale."

VAT

A further factor to be considered is the effect of VAT on building costs. New buildings do not carry VAT, but building materials do. A builder can reclaim the VAT which he pays on materials as a monthly routine, but the private individual cannot do this. Fortunately, there are special arrangements for self-builders to reclaim VAT, and the relevant circular, VAT Notice No. 719, is reproduced at page 167. The procedure is simple, and the authorities concerned are most helpful. The effect is that whether you use a builder or build for yourself, VAT will not be included in your finished costs. However, you cannot make your claim until the building is completed. It will then be paid fairly promptly. Your total claim will be well over £1,000 and will have to be financed in your building budget. Fortunately, VAT is not payable on sub-contractor's services. Farmers and other VAT registered self-builders can reclaim VAT through their normal farm VAT return if the building is being funded through their business accounts.

NHBC Certificates

The National House Builders Council is an organisation to which virtually all house builders belong. It concerns itself with housing standards in many ways, and houses built by builders registered with the Council carry NHBC certificates which are virtually ten-year guarantees on a property. These offer invaluable protection to the home-buyer, and also give some protection if the builder goes bankrupt before he finishes the job. Most people with their own land seeking a builder to build for them restrict their enquiries to NHBC registered builders, and these builders are obliged to arrange for an NHBC certificate on every house which they build which is not architect supervised. The NHBC has inspectors who pay routine visits to the property, and who call in the NHBC civil engineers and others if foundation or other problems are encountered. If any work is not up to standard the Council will insist that it is rectified, and if this is not done the builder will lose his registration. A contract with a registered builder should always include an arrangement for an 'NHBC form 5C agreement'. Full details of this are found in the leaflets which the builder will provide, or which can be obtained from the Council at the address given in appendix 3.

Unfortunately there is no way that a self-builder can obtain NHBC membership, and it is unlikely that any NHBC member could be induced to act as the 'nominal' NHBC builder for a self build project. Not only does this mean that no self build home can have the NHBC ten-year certificate, but it means that the self-builder has to offer his building society some other form of guarantee. Normally this is by means of architect's or surveyor's progress certificates. Some building societies, notably the Nationwide, have their own surveyors, but most will expect the mortgagee to find his own qualified man to undertake this work. If an architect has been retained, he will undertake it, and similarly the package deal companies will also make appropriate arrangements. The self-builder who has drawn his own plans, or who has used an unqualified designer, will be at a disadvantage here, and this single factor may completely rule out building societies as a source of finance — or rule out the unqualified designer.

All the materials for a new home. The photograph was posed for The Sunday Times by D & M Ltd., and shows their clients with the author, the engineer, the materials manager, and everyone else involved in providing the D & M service. In the background are the materials supplied—a giant Lego set.

Buying a Site

Owning a piece of land involves far more than simply paying for it and moving onto the site. The business of buying or selling a building plot has its own vocabulary, and its own set of procedures. Solicitors, land agents, and estate agents make their living by operating this system, and do not go out of their way to explain what they are doing to their clients. If you are to buy land it is important to accept that you are going to have to accept the system, and learn the vocabulary. Property titles *can* be conveyed by private individuals. Money can be borrowed with land as a security without a legal charge or other documentation. Easements, access rights and other matters can be dealt with by those who are not solicitors. Depending on your viewpoint this can be worthwhile, or a recipe for a legal disaster. For the self-builder the savings obtained are simply not cost effective. The time and effort spent in transferring land without a solicitor is disproportionate to any saving, and the suspicion with which your title will be regarded when you use the land in some way, as security for building finance for example, can be a nuisance. If you are looking for finance you are loading the odds against yourself with a home-made title to your principle security.

If you already have, or know, a solicitor who you wish to act for you, that is fine. Otherwise you will probably entrust your affairs to the man whose office you first contact. If you are buying a plot of land with planning consent it is unlikely that there will be anything to choose between the competence with which different firms of solicitors will deal with the purchase, but there may be a big difference in the speed with which they will do the job and the advice which they will give to you as a self-builder. If time is important it can speed things up to use a local solicitor as he will be familiar with local titles. The only way to be sure that you get good advice is to rely on recommendations. Explain from the outset that you are building for yourself and that you may be looking for advice on various matters. Explain your timetable and ask if it is practicable. Ask about fees. With luck the solicitor will take a special interest in an unusual client and be of considerable help. If you do not establish a rapport you can ask outright if he really thinks you are the right client for him.

Always have a word with your solicitor once you have decided to buy a piece of land, but before you are committed to it. He will advise on how to make your offer, and will make sure that you are not irrevocably committed before he has a chance to examine the title. This normally involves signing papers only if they are qualified by the words 'subject to contract'. This means that you can back out at any time until you sign a contract. However, if you buy at auction, once the hammer has fallen you have 'bought it', snags and all, and there is no going back. If going to an auction consult your solicitor beforehand.

If the land you want to buy is being sold by an estate agent he will probably ask you to sign a note and pay a deposit. This is quite usual, provided that the words 'subject to contract' appear, and that the deposit is refundable if you do not proceed. If you are negotiating a private purchase you may wish to make a written offer, or to accept an offer made by the vendor. Your solicitor will prefer to write these letters for you. Alternatively, you may simply shake hands with someone on an agreed deal, with no paperwork involved, in which case all you need to know is the name of the vendor's solicitor.

The actual purchase of the land is started by the vendor's solicitor sending your solicitor two copies of a 'draft contract' or agreement to buy the land. This is unsigned and comes with details of the title. Your solicitor will look at this with a critical eye and initiate various standard procedures to make sure that you really know what you are getting. Many of these are called searches, and include asking the local authority to confirm that it is not aware of any orders relating to the land, that it is not about to be compulsory purchased and confirming any existing planning consents. This takes time, and little can be done to hasten searches. When they are finished you will probably meet your solicitor and he will explain the facts he has established.

At this point it will help to produce your own site plan, with your own notes of the access, drainage, water supply, electricity and gas supply that you will require. Ask your solicitor to confirm that this is what you are buying. Tell him the sort of property that you want to build and ask if there is anything in the title to preclude this, such as covenants which restrict the height of any building on the land.

All this refers to a single plot with planning consent — either outline planning consent or a full consent. At this stage, while you can still back out, you will be looking into exactly what you can build under the terms of the consent. This is a job for your architect or whoever is handling your application, and not the solicitor. Your architect will now probably submit your planning application for you, which you are legally entitled to do as a prospective purchaser.

Eventually you will reach a stage where your solicitor has satisfied himself that you are proposing to buy a valid title, and has explained to you exactly what rights and obligations are involved. You will have obtained your own planning consent or are satisfied that you are going to get it, or are happy to purchase with the existing consent. It is important to differentiate between these situations, which are that either

1. the land has an outline consent and you have obtained approval of the reserved matters in that consent to build your dream house.
2. OR the only issues left over the reserved matters concern details which you know will be resolved and you will go ahead anyway.
3. OR you are buying with outline consent and will submit your detail design later, knowing that the planners will have to be convinced that it is the right design for the site.
4. OR you are buying with a full consent and will build to the design for which that consent was granted.
5. OR you are buying with a previous full consent and have obtained your own subsequent full consent or a letter agreeing to a variation of the original consent design.
6. OR you are buying with a full consent and are confident of obtaining consent for your own design which you have started to negotiate with the planners.
7. OR you are buying with a full consent with the intention of getting the design varied after you have bought.

If your situation does not fall into one of these categories, you are involved in a special situation, and require special advice. Having cleared all this you sign the contract, which has already been signed by the vendor, and pay a proportion of the agreed price. You are now committed to the transaction. If you try to pull out the consequences are expensive. It is essential that the money to buy the land is to hand before you sign a contract, and if it is not in your bank account, you should discuss the way in which it is going to get there with your solicitor. He will take a jaundiced view of a client signing a contract with the backing of a great-aunt who has promised to produce the money when the time comes.

When you have signed a contract you have only contracted to buy the land, but have not yet bought it. Technically you may not use it, but your solicitor will probably be able to arrange with the vendor for you to make a start, particularly if you have already given him the full purchase price and he is able to assure the vendor's solicitors that he has the money.

At this stage consider any insurances appropriate to your new situation as a landowner. If you are starting work at once then your building insurances will cover all hazards, but if there is a delay then it is wise to have insurance to cover any claims, however unlikely, from trespassers who come to harm on your land, or a passer-by on whom your tree happens to fall. The cost is negligible.

The contract will require that the conveyance of the land shall be completed within a set period, usually a month. This is known as 'completion' and involves your solicitor drawing up an unintelligible document which is signed by the vendor and delivered to your solicitor in exchange for the balance of the purchase price. This final part of the transaction is something of an anti-climax. The actual title deeds, which essentially comprise all the conveyance documents relating to the land in date order, finishing with the one conveying it to you, are usually sent straight to the bank as security for your building finance, and you never even see them.

Still, you are now a landowner and it is high time you became a builder.

A typical Milton Keynes owner built home. The Milton Keynes Development Corporation makes plots available to self-builders on a 'build now, pay later' basis, and on any weekend you will see dozens of families working at building their own homes there.

Design

Among the many illusions concerning self-build projects is the idea that those building for themselves are able to express their own personalities by building homes that are uniquely their own, and that they can indulge their wildest dreams. In fact the self-builder suffers many of the same constraints as the developer, and if he avoids standardised features it will only be with considerable effort. All influences in modern society urge towards conformity. Many features of this conformity are desirable, and reflect the best in modern living standards, but others are imposed, directly or indirectly, by the planners, the dictates of finance, what is practicable in a given situation and, if housing is being built by a group, the views of that group.

Groups have to ensure success, and they do this by avoiding anything which complicates the already complex building process. This is particularly true in the field of design, whether in the layout of the site, or in the design of the houses themselves. In practice 99% of Housing Association self-build homes are three bedroom semi-detached houses or four bedroom detached houses, always of conventional design. Site layouts are invariably attractive and practicable, but never trendy.

For all self-builders the budget determines the design. Invariably the aim is to build as large a *conventional building* as possible within the budget, which means the lowest cost per square foot that can be achieved using *conventional construction methods*. These two qualifications are important, for although revolutionary building techniques and experimental designs are sometimes claimed to offer very low costs, they invariably present other difficulties, particularly over mortage valuations. Revolutionary construction techniques make good media copy— but nothing else.

The low costs goal is most easily achieved if the following features are acceptable. Firstly, a level site and a solid ground floor are most important. A split level building or a ground floor built up above surrounding land on a sloping site can add two or three pounds per square foot to the cost of the most simple bungalow. This does not rule out a sloping site, but it requires the buildings to be built on terraces cut into the slope, rather than built out from the slope. Imaginative landscaping can make the two approaches look much the same, but the costs are very different.

Similarly, the shape of the building above the ground should be simple. The cheapest shape to build is a rectangle with two end gables. 'L' shaped buildings, gable projections, hipped roofs, all increase costs. Dormer bungalows have lost their attraction due to the rise in the cost of timber. The ideal is the simplest possible box with the simplest possible roof on top of it. This simplicity of shape need not impede an attractive final design which will depend on proportion, on the type and spacing of windows, and on the detailing of chimneys, porches and other features. Where it is necessary to add extra features to a simple and economical design, this is easily done with landscaping detail which can be effective and economical. Providing a link wall between garage and house, a feature chimney, a buttress adjacent to a porch cost little but can add a lot to the character of a building. Good design is a matter of simplicity done well and if costs have to be held down, the simplicity is essential.

Unless one has experience in designing houses it is certainly not a field for Do-It-Yourself. Housing Associations find that they have to make use of architects, or a recognised design service, and those building on their own should do likewise. The difference between architects and the others are important.

Persons may only call themselves architects if they are fully qualified members of the profession, registered with the Architects Registration Council and practising in conformity with its rules. If they advertise they may only do so discreetly, and should charge fees on a set scale. By using an architect one is assured of work to a high standard with safeguards against any loss due to professional negligence. The Royal Institute of British Architects publish various booklets for the public on making use of the services of their members, which are available from them or most practices. The cost of an architect's full service for an average house starts at 7% of cost plus incidental expenses.

Designers offer the same service as architects but with limitations. They operate outside the umbrella of the R.I.B.A. and usually lack formal architectural training, although they may be qualified design draftsmen. They vary in size from established companies to employed individuals who prepare plans for private clients in their spare time. A large proportion of this latter category are local authority building inspectors and other technical staff, and this is generally recognised by their employers on the understanding that they do not prepare drawings for submission to the council which employs them. The fees charged for the services of designers who are not qualified architects are open to negotiation, and their clients do not enjoy the safeguards extended by the R.I.B.A. to those who employ its members.

Companies who offer a design service normally work in a different way, and can be divided into two groups, those which sell their design services as part of a package supplying the materials for the building itself and those which sell only plans for a range of standard designs. The latter vary from concerns whose principle interest appears to be to sell their books of plans, selling the drawings offered in the book only if pressed to do so, through to well known companies. The services they offer are highly competitive in price. At the prices charged for a set of plans the service offered has to be formalised, and clients cannot expect any more than is set out on their order form. The designs which they offer are drawn to have consumer appeal rather

Regional design — The planners now expect new houses to be built in local materials in a regional style. This new farmhouse is in Lincolnshire.

than low construction costs. The drawings are always competant, as they have to be to conform to Building Regulation requirements.

Using the design services offered by the package companies must depend on a decision to use this approach to building. If one opts for timber frame or concrete panel construction then the drawings have to come from the manufacturer and the nature of the building system will restrict the opportunity to have a standard design altered beyond set limits. A package of traditional materials, however, usually enables the standard designs to be varied to almost any extent, or for drawings to be prepared to a client's own requirements. All the building system and package companies will quote firm prices for their materials and services, and give realistic figures for the probable cost of the finished building. In deciding between them this is always useful.

Before considering a design for a new house the site for it must be found. This may seem obvious, but many people looking for a site have already decided what their ideal home should look like and are looking for somewhere to build it. This approach, although understandable, is also illogical in a country where any building site is at a premium, and where there is no shortage of professionals to design dream houses to suit any situation. Everything starts with the site, which will determine the most basic features of the design. The decision whether the windows in the main rooms should be on the same side as the main entrance can only be settled for a specific site with a specific view, specific access, and specific orientation. Until it has been found it is more important to make a note of design features that appeal rather than to get involved in a drawing.

Obtaining a Design Study

When one has a site, and a decision whether to approach an architect or a design service company has been made, the next step is to obtain a design study. This is a simple drawing which shows a proposed design in sufficient detail for it to be the basis of full design drawings. A selection are shown on pages 36 to 41. A design study is meant to be discussed, altered and changed until it shows the home that the client wants. Then, hopefully there will be no more changes. It has to show the building on its site, so that distances from boundaries, access, shape of garden and many other matters can be considered and it usually incorporates a perspective sketch. The designs in the books of plans published by the design firms are design studies without site plans.

It is not usual for architects to suggest to clients that a design study should be the first stage in a commission. This is because an architect does not want the client to ask himself "do I want to use this architect's services if this is how he interprets my requirements?" He would far rather keep the arrangement moving gently along, until it reaches a compromise between his professional advice and the client's ideas, and he is established firmly as the client's professional agent. Maintaining this role is difficult because the client, wanting a modest house within a fixed budget, feels that he wants a clear idea of what he is going to get before he commits himself. Also, at some point in the design process, costs have to be quantified, and the decision made that the design proposed is practicable within the budget. An agreed fee for a design study enables this to be done, and avoids the risk of a design developing to full working drawings, and a planning application, which the client cannot afford. This may seem ridiculous but it often happens and is rarely the fault of the architects. It usually stems from the client not telling the architect exactly what he can afford, and from both parties avoiding the ungentlemanly topic of money in their discussion. If money is a key factor in the whole business it should be treated as such, and at an early stage the client should insist on costs

Regional Design—Houses in Herefordshire which the planners required to mimic the traditional cottages in the village.

being discussed. It is also reasonable to ask if the architect will undertake this work at a fixed price.

Established practices will not like to be asked how much design a fiver will buy, but an enquiry explaining that one is seeking preliminary sketch designs for a new home to enable budget costs to be considered, and enquiring what the cost of such a design study would be, indicates that you are likely to be a realistic client. The reply may be that the practice prefers to be retained on a firm basis, in which case it is probably so successful that it does not touch anything in your price range at all. The reply is more likely to suggest a meeting at which a figure of between £80 and £150 to cover the study will be proposed. It may equally lead to a phone call from a newly qualified architect offering to come and talk about what he can do, and to show you preliminary sketches absolutely free. Whatever happens, the prospective self-builder should explain his intentions to the architect, and should accept that it is wrong to commission a design study, decline to go further with him, and then have the same design drawn out in detail by someone else.

In practice, few individual self-builders use qualified architects, and unless they use a package deal service, they invariably go to part time housing designers. Here the same advice applies—commission a study at an agreed price on the understanding that you will go no further unless it is likely that it meets your budget. Some moonlighting design draftsmen will charge as little as £50 for the sketch design required.

With the package deal companies, or those selling sets of plans, things are more cut and dried. They all publish details of their standard designs, and set out clearly the commercial basis of their arrangements with clients. If designs can be modified, or drawn to a client's own requirements, this is detailed at length. Most have experienced field staff who call on enquirers without obligation, and will arrange site surveys and other work for a fixed fee. Above all, they offer prospective clients the opportunity to see show houses or other completed buildings for which they have been responsible. They do not accept all the responsibilities borne by architects, and they do sell standardised designs which may or may not suit an individual's requirements. For the average self-builder they have one advantage which may not at first be obvious. As commercial operations, with no professional trappings, it is easy to reject their services after a preliminary meeting. The service they offer varies in competence, but with well established companies this only varies from adequate to excellent as the need to get drawings approved under the building regulations filters out the substandard. Even so, it is worth asking if drawings are prepared under the supervision of qualified staff architects. A staff archiect is important, as a consultant architect to a package deal company may not handle the affairs of individual clients. If possible you want the signature of a qualified man on the bottom of your drawing.

Successful design is a compromise between many different requirements. A design has to meet your budget and accommodation requirements, to have a style and appearance acceptable both to you and the planners, to meet the building regulations, to take best advantage of site features and existing site services, and to maximise the site value so that the house design built is the best possible investment. Your role as the client in the design process is to define your budget, accommodation requirements and stylistic preferences clearly and unequivocally. Avoid sketches unless you have experience as a draftsman. We are all experts at communicating with words, and will convey our ideas to a designer better this way than in a drawing. Your sketch designs may also restrict the professional.

Briefing a designer, or selecting a standard design from a book of plans, you must consider exactly how you want to live, and exactly how you want a

The Eighties cottage look: self build house under construction on a Northampton development Corporation site.

The Seventies Look. The popular ranch style bungalow is now only likely to get approval in suburban areas.

home to look. Often the question of style and appearance can be explained by reference to an existing building, photograph or illustration. Building for yourself provides an opportunity to build to suit yourself, and time taken to analyse what you want from a building is seldom wasted. On page 35 is a typical checklist of design requirements which one architect puts to his clients, and which has been the basis of hundreds of individual designs. It is concerned only with features which affect the shape and layout of rooms, and thus of the building, and not with fittings and fixtures in them. However, before studying it, it is worth looking at what sort of accommodation can be provided within a given area.

The area of a building is normally taken as the area enclosed by the internal faces of the external walls, which is considerably less than the overall area covered by the structure. All areas in this book are measured in this way, as are the areas in most books of plans, and are expressed in square feet or square metres. For practical purposes 11 sq. ft. equals 1 sq. m. The best way to examine just what can be provided in buildings of different areas is to study books of plans showing the areas as well as the room sizes on each plan. The following are the characteristics of average houses and bungalows of different sizes. Remember the areas quoted are the areas enclosed by the external walls, and not the overall plinth areas.

Up to 700 sq. ft.
Holiday chalets and 1/2 bedroomed old people's bungalows only.
700 to 800 sq. ft.
Smallest possible three bedroom semi-detached houses.
Small two bedroom bungalows.
800 to 900 sq. ft.
Small three bedroom bungalows with integral lounge-dining rooms and compact kitchen.
Two bedroomed bungalows with larger kitchens or a separate dining room.
Most estate built three bedroom semi-detached houses.
Around 1000 sq. ft.
Large three bedroom semi-detached houses.
Three bedroom detached houses.
Small four bedroom houses.
Four bedroom bungalows with integral lounge/dining room.
Three bedroom bungalows with separate dining room or large kitchen.
Luxury two bedroom bungalows.
Around 1300 sq. ft.
Three or four bedroom detached houses and bungalows with the possibility of a small study, or second bathroom, or a utility room, or a very large lounge.
Around 1600 sq. ft.
Four bedroom houses or bungalows with two bathrooms, large lounges, small studies, utility rooms.
Around 2000 sq. ft.
Large four to five bedroom houses and bungalows.

Central heating and a high level of insulation are now virtually standard in new construction. Building regulations are calling for higher levels of roof and wall insulation, and it is probable that this will extend to floor insulation in due course. If a large window is to be a feature of a wall it must be double glazed to achieve mandatory insulation levels; otherwise double glazing is a matter of cost and personal choice. Central heating does not save energy, but it does heat a house in the most economical way, and without it a new house is unusual and this may affect its resale value. If it cannot be afforded to start with you can make provision for it in the structure, and if possible piping or duct work should be provided even if the heating unit cannot be installed.

So far we have looked at design and the set patterns which tend to emerge to meet specific criteria, but in building for yourself this need not be the limitation that it appears to be. The constraints of planning and finance affect only the structure of the building, its shape and how it is built. There is far more to a house than this and if you think about homes which you particularly admire it is interesting how much of their appeal comes from their fittings, decor and landscaping. Here the self-builder can give full rein to his imagination and can achieve most reasonable ambitions without straining his budget. It is surprising how little it costs to achieve special effects when these are planned in advance and provision for them is built into the structure as work progresses. Obviously some luxuries are costly—circular baths for instance, or marble fireplaces, but others are cheap, and can add enormously to the total effect. An elaborate electrical installation costs very little if installed before the house is plastered and, by imaginative wiring, luxurious effects can be achieved using standard fittings. Recessed ceiling lights, pelmet lights, outside power points on a veranda, wiring for a garden floodlight all cost little more than conventional lighting. The cost of an additional light or power point is about £10. Lights inside cupboards that turn on automatically when the doors are opened need cost no more than the cupboard door knobs. T.V. aerial points can be provided in every bedroom for no more than the cost of the wire and the socket, and provision made for a plethora of telephone extensions. Built in hi-fi wiring is very cheap indeed. Even if the cost of the fittings is unacceptable it is possible to provide wiring of all types for future use. Before plastering it costs virtually nothing, and will be hidden; after plastering any new wiring is expensive and ugly.

If an additional bathroom or w.c. might be required later, it is simple to make provision for this in the drainage system. Recesses in walls for built-in wardrobes are common-place, and are also practicable for bedheads, sideboards, drinks cupboards and bookcases. Architectural niches with frosted glass shelves that conceal lights to illuminate an ornament or flower arrangement from below are cheaper than any painting if built in as work progresses. The same applies to recessed mat wells at a front door, or fireplaces built against an outside wall with monster ashbuckets that hold a week's ashes and are removed through a trap outside the house. Special provision can be made for hobbies, such as aquaria built into walls, or display shelves. Wall safes can be built into walls or floors. Power operated garage doors worked by remote radio transmitters carried in a car can cost less than a washing machine. All these features are expensive to build into an existing house, but if planned early it is possible to provide them at a realistic cost. This is one of the hidden benefits of building for oneself.

In planning bathrooms and kitchens the opportunities for an individual layout are obvious, and offer the bonus that a striking design will often permit relatively cheap fittings to be used. The best examples are to be found in kitchen supplements or bathroom design features in women's magazines, for commercial displays invariably emphasise the more expensive fittings on sale. A luxury kitchen or bathroom in a new house need involve only a fraction of the cost of luxury fittings or sanitary ware. If a bathroom has bathroom cupboards recessed flush into the walls, imaginative concealed lighting and the ceiling lowered over a recessed bath to provide fixing for a shower curtain, it can be fitted with standard white sanitary ware to set off well chosen tiles—and all cost less than a medium priced suite of sanitary ware in a fashionable colour. The opportunities to do the same in the kitchen are even more exciting, and many of the interesting layouts detailed in books of kitchen designs are based on cheap contract grade units used in an imaginative way, often with a custom made worktop above. Again, special lights can be used to great effect, and wall tiles in interesting patterns and colours cost little more than the commonplace.

In landscaping the new house the self-builder can really express himself as planning consents rarely concern themselves with garden arrangements or with garden walls. It provides the opportunity to dress up a simple building or to emphasise its simplicity. It can be used to make a small building appear larger, or to give perspective to a large building that would otherwise overfill its site. This subject needs as much care and consideration as any other aspect of design. Gardens, although made by years of thought and work, have their basic layout determined at the time the house is built and it is not easy to alter them later. Apart from the house itself, the dominant features in most gardens are the garage and the drive. If the building plot is small, or the frontage is restricted, there is little choice of layout, but when this is not the case the relationship of the garage to the house, and the shape of the drive, should be dictated by aesthetic as well as practical considerations. It is a condition of planning consents that a vehicle turning area is provided within the plot, and often this turning area can become a garden feature, limiting it with dwarf walls or a rockery, or building steps or a path to provide access from it to the garden. A wall linking a house and its garage, built in the same brick to match the house, has the effect of making the two simple structures appear to be a far larger complex. Concealing the back garden in this way usually makes it appear larger than it really is. Garden layouts are beyond the scope of this book, but remember that many features of the house will determine what is possible for the garden, and that many garden features can be provided easily while building work is in progress.

Regional Design—Bradshaw Housing Association house in W. Yorkshire.

HOUSE/BUNGALOW LAYOUT—DESIGN CHECK LIST

Front Entrance	Do you think an impressive front entrance should be a key feature of the house? Or just take its place in the front elevation?
Front Porch	Is this required? Is a storm porch required to draught-proof the hall when the door is opened?
Front Hall	Is it important that the hall should convey the whole feeling of the house as soon as you step inside? Or is it just a space between other rooms? What furniture do you want in the hall? How essential is a cloaks cupboard? Do you prefer a glazed or solid front door?
Lounge	Lounge or lounge/dining room? Is the feel to be of an enclosed room, with no more window than to provide a view, or is the preference for a large area of glazing with the room relating to the garden or patio outside? Are patio windows required? Is a fireplace required? If not, perhaps a dummy fireplace? If so, a feature fireplace or a classical small fireplace? If a feature fireplace, should this have a vertical feel, to the ceiling, or be a full wall feature, or have an extended mantel to provide shelving? If you are having a lounge-dining room, is some form of room divider or natural break between the two parts of the room preferable? Does the lounge/dining room need a door to the kitchen or just a hatch? In general terms, do you like 'L' shaped lounge/dining rooms?
Dining Room	Maximum number to sit at table? Door or hatch to kitchen? Is a built-in sideboard acceptable to save space?
Kitchen	Will the family eat in the kitchen? At a breakfast bar or table? How many? What major appliances are required in this room? Is a structural larder required? Is an Aga or similar solid fuel cooker required?
Utility Room	Need this be any more than a large porch? What appliances are required in it? What storage space? A w.c. adjacent to the back door (useful in rural areas)?
Bathroom	One bathroom or two? If two, is one to be en suite with the master bedroom? Fittings in bathrooms—bath? basin? w.c.? bidet? separate shower? Airing cupboard in bathroom, or can it be elsewhere?
W.C.	Where?—in bathroom(s)? in cloakroom or hall? at back entrance? Basins—in which w.c.'s? Cloaks cupboard—in a cloakroom which is also a w.c., or would you save space by having hooks and rails actually in the cloakroom itself, not enclosed in a separate cupboard?
Bedrooms	How many double, how many single? Is the master bedroom to be as generous as possible at the expense of the others? Is provision to be made for built-in furniture?
Study	Is this to be a significant room, or very small with just room for a desk and filing cabinet? Does it need built-in shelving, or cupboards? Will it double as an occasional bedroom, and thus require room for a divan?
Garage	Integral with the house with a communicating door to the utility room or kitchen, or separate? How many cars? Plus extra room for garden tools etc.?
Central Heating	What type of central heating system is envisaged?
Future	Is there any possibilty of future extensions or alterations to suit changed family circumstances?

TYPICAL DESIGN STUDY

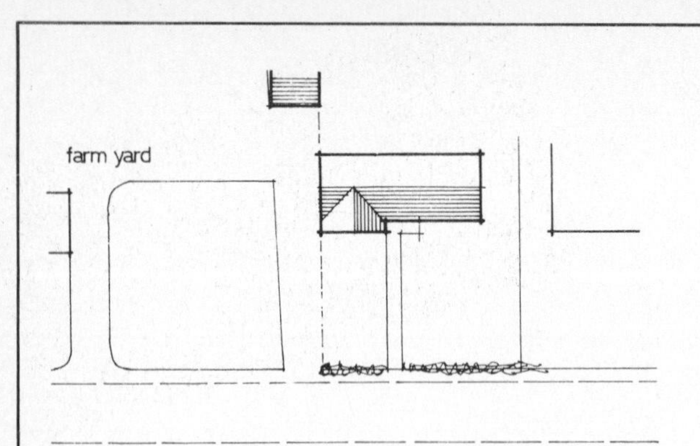

Site Plan 1:500

A design study is the first stage of any new home, and is used to settle basic design features and to establish budget costs. Do not get involved in more elaborate drawings unless the design study shows exactly the new home you want, and it is confirmed that it can be built within your budget.

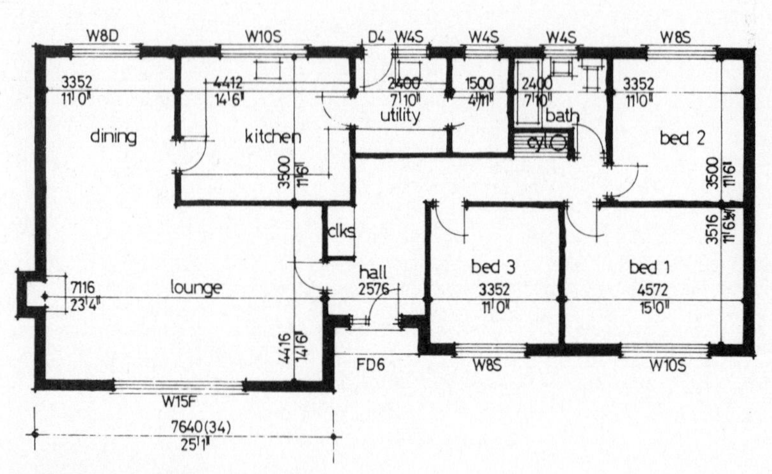

PLAN 1:50

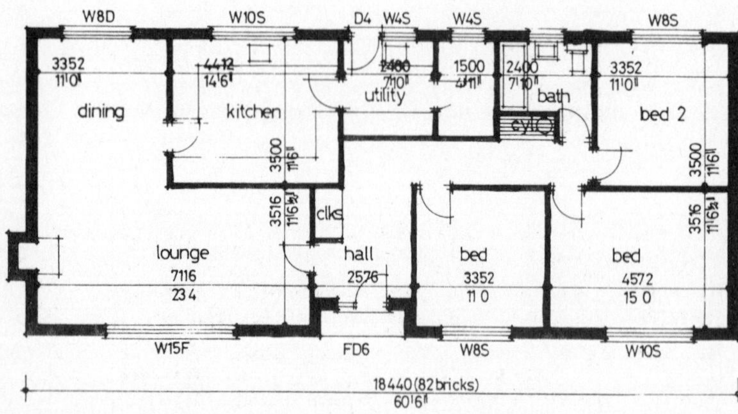

Alternative PLAN 1:50

Another design study for a compact bungalow, including a site plan. This is an actual study drawn up by a designer who is not a qualified architect: the presentation lacks polish but the essentials are all there.

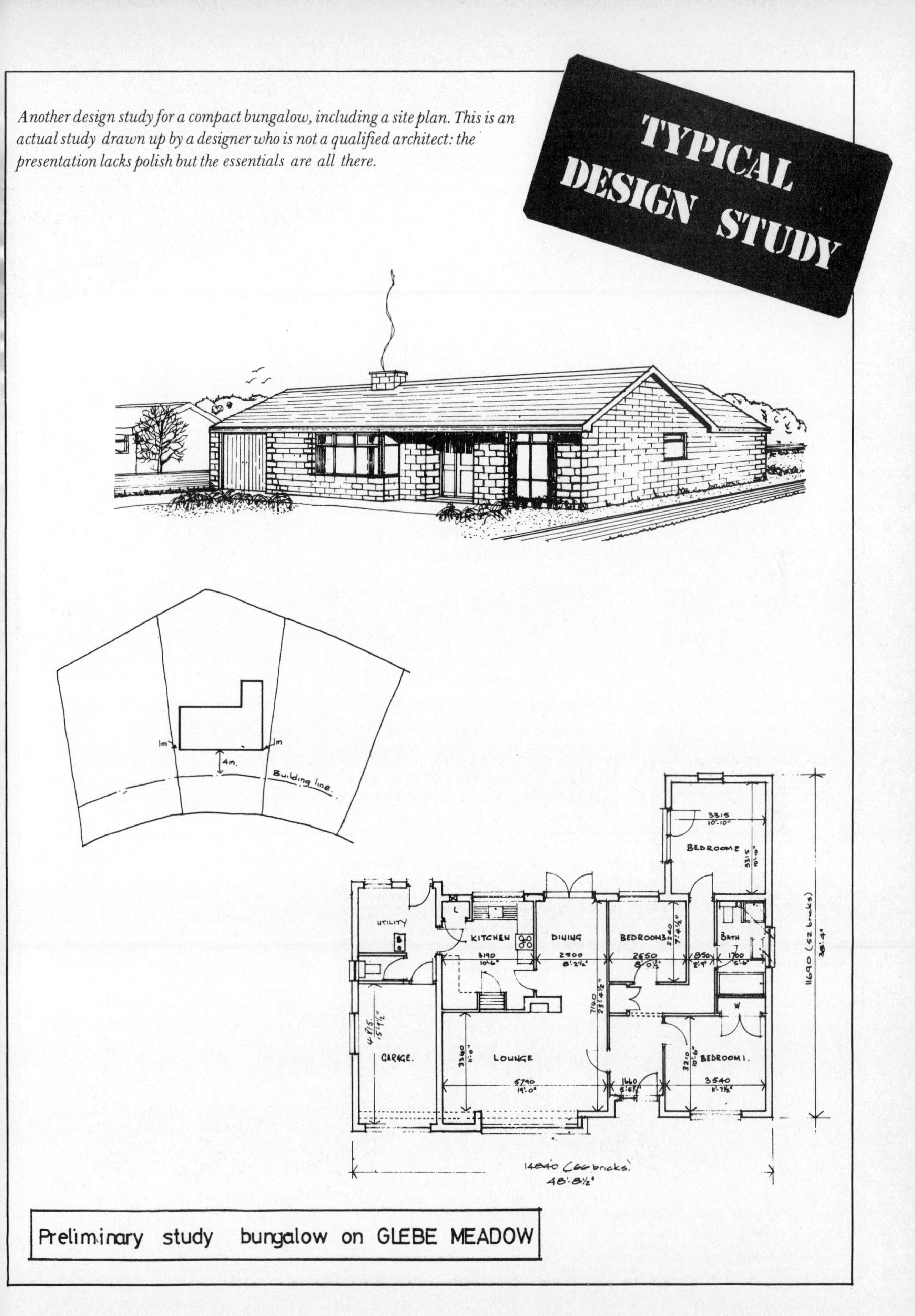

158	sq m
1700	sq ft
368	cu m
13032	cu ft

This design study for a large ranch-style bungalow gives plenty of plan detail, but lacks a site plan or any indication of room heights.

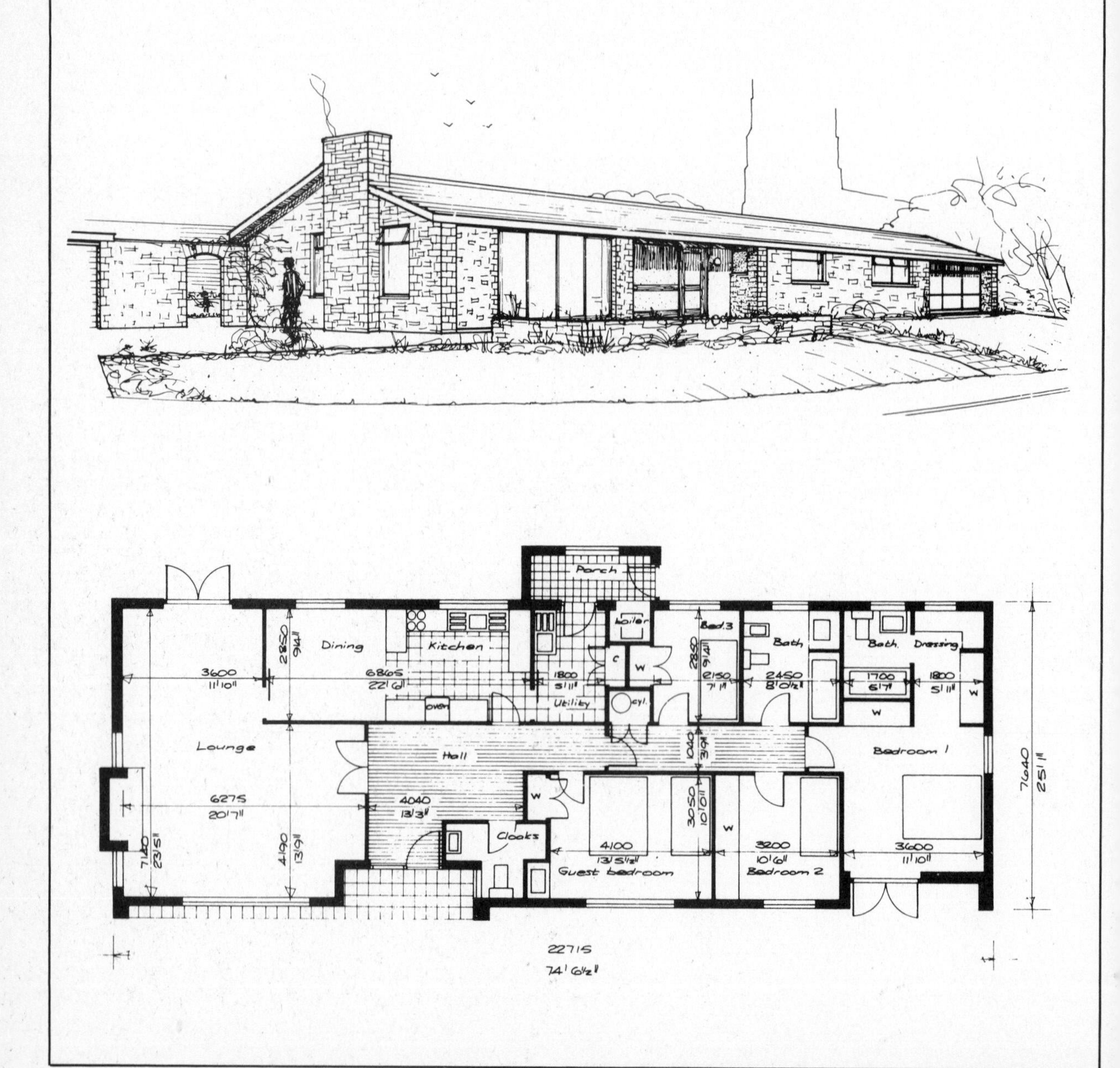

Preliminary sketch for a very large bungalow, in a distinctive style to suit a very special site in a Cambridgeshire village. There is not enough detail here for any but the roughest estimate of costs, and it is drawn as the first of a series of studies.

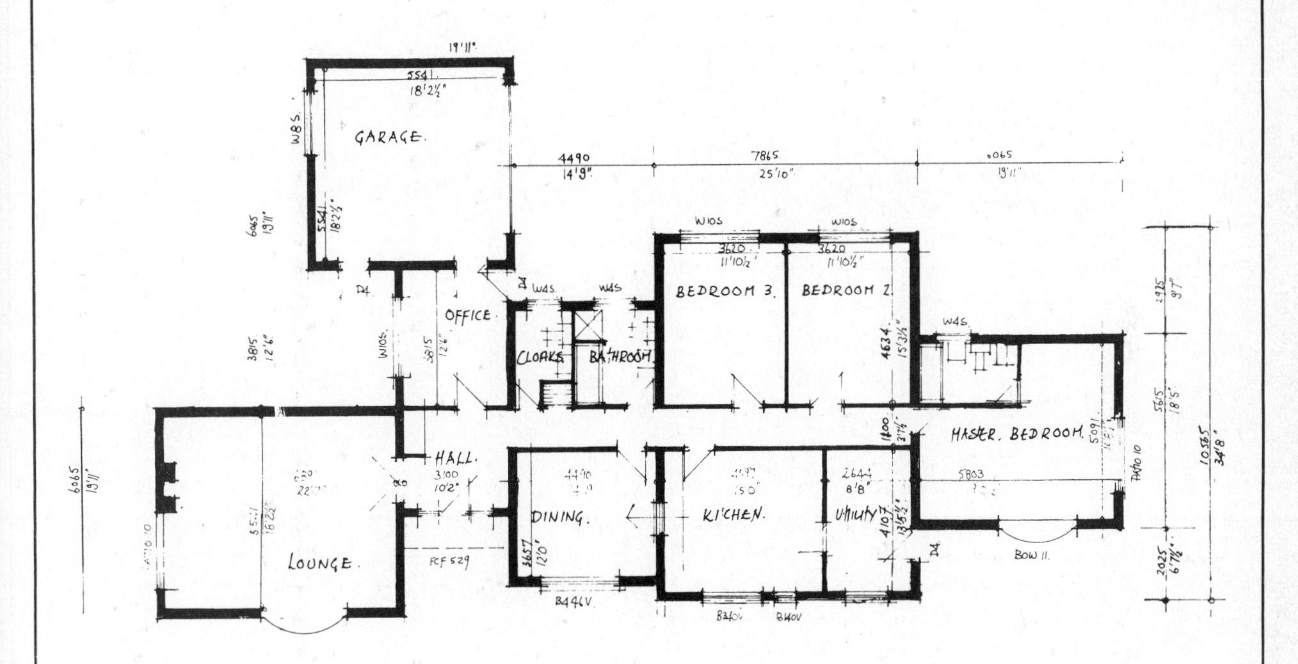

PRELIMINARY SKETCH FOR CLIENTS APPROVAL

DRWG No 2176/A/Prelim

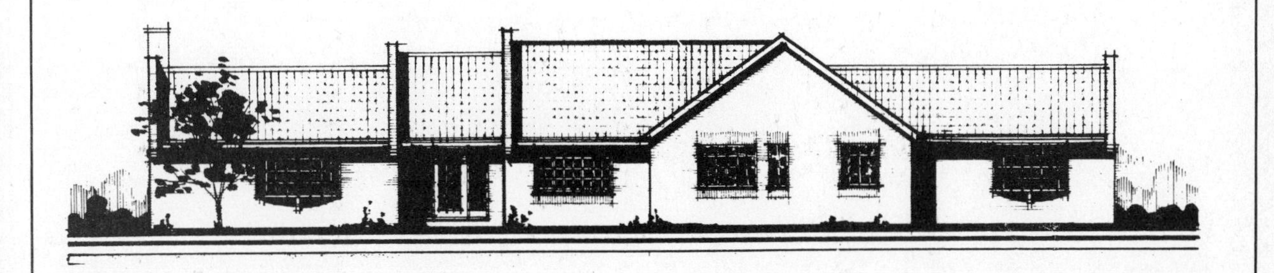

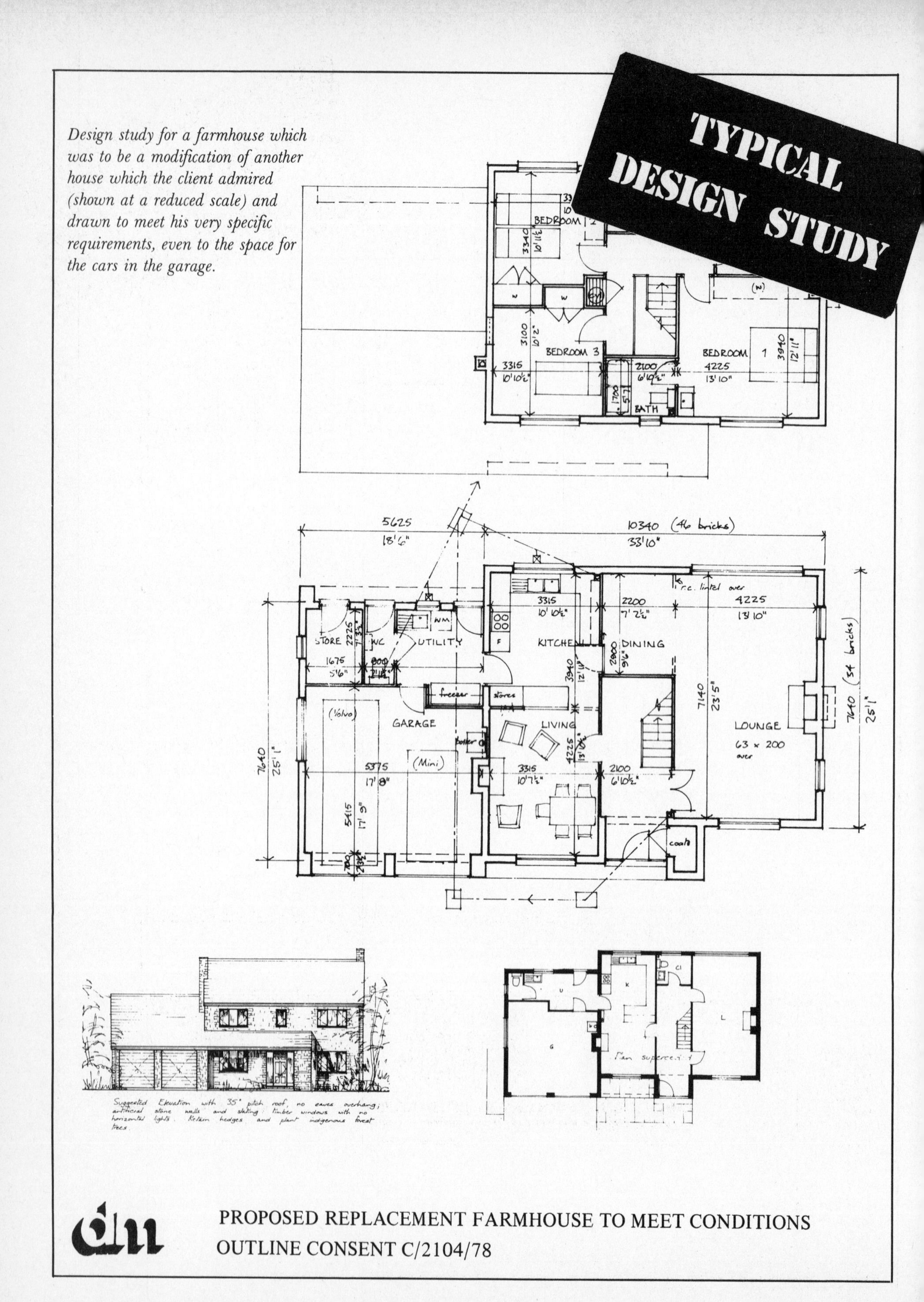

Design study for a farmhouse which was to be a modification of another house which the client admired (shown at a reduced scale) and drawn to meet his very specific requirements, even to the space for the cars in the garage.

PROPOSED REPLACEMENT FARMHOUSE TO MEET CONDITIONS
OUTLINE CONSENT C/2104/78

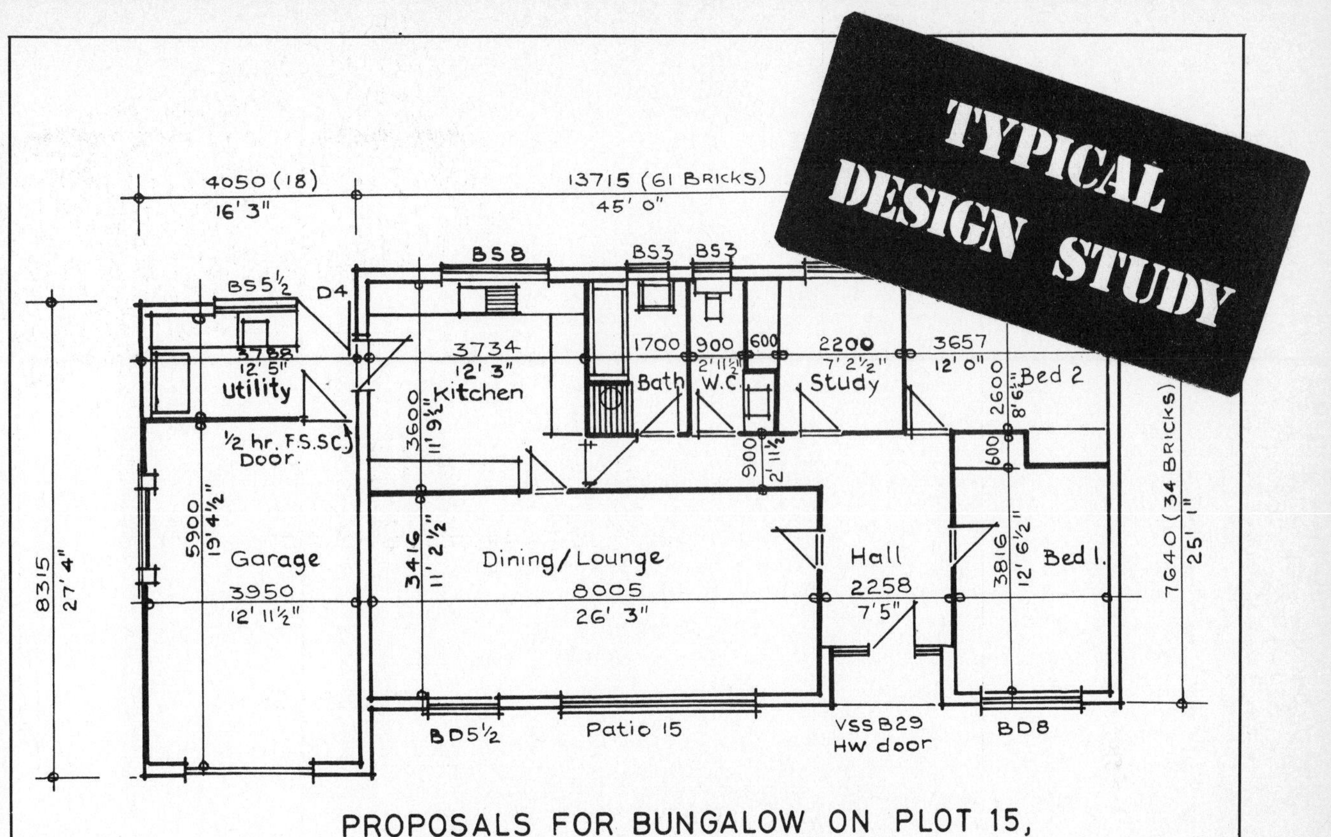

PROPOSALS FOR BUNGALOW ON PLOT 15, ELM CAUSEWAY DEVELOPMENT.

The Architect who drew this sketch for a simple rectangular bungalow has shown a strict elevation instead of a perspective sketch. It is a felt-pen drawing, and this severe style of presentation takes some getting used to.

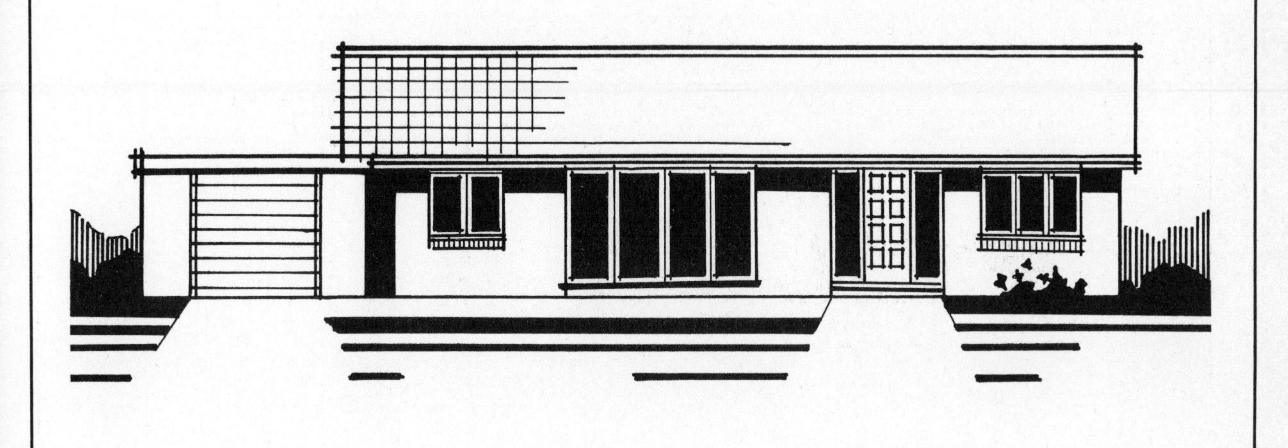

Drawings and Specifications

Most medieval cathedrals were built with the aid of fewer drawings than are now considered essential for a municipal public convenience, and many self-builders follow the former example. However, formal drawings are essential to obtain Planning and Building Regulation consent, and if a full set of construction drawings is not going to be prepared for use on a site, the decision to dispense with them should be a conscious one, taken with a clear understanding of the options.

The drawings required by a Local Authority must include:

A location plan to a recognised scale to identify the site.
A site plan, normally to a scale of 1/500, to show all boundaries, the position and siting of the building, drainage arrangements, access etc..
A building plan at a scale of not less than 1/100, together with at least two elevations of the building.

This detail can be presented on a single sheet, or in any practicable way provided that it is to the appropriate scale. At least six copies are required, and some authorities require one printed on linen, meaning a cloth reinforced paper. If alterations or additional information are required by the authority, then a further set of prints will be needed. To facilitate this, drawings are always prepared on tracing paper or tracing film, from which prints are made by the dyeline printing process. Most towns have document reproduction centres where copies of drawings can be made while one waits. The cost is quite small. The transparent master drawings are easily altered without having to be redrawn. It is usual for one copy of the approved drawing to be returned with the Planning and Building Regulation Consent certificate, with the approval stamp endorsed on it. This approved drawing must be followed when building, and any subsequent alterations agreed with the authority should either be shown on a revised drawing sent to the Local Authority for approval, or else clearly detailed in an exchange of letters. It is important to understand that the details shown on the drawing which has been approved under the Building Regulations are only particular aspects of the work which it is considered should be specially emphasised. All relevant requirements of the Building Regulations have to be complied with, not only those shown on the approved drawing.

The construction drawings for site use may be the same as the ones submitted to the council, or may be drawn separately. If the latter, they must contain all the details shown on the approved drawings and remember that they will be used by building tradesmen and sub-contractors as well as by the self-builder himself. There are many advantages in using fully detailed drawings. They can save a lot of argument if a problem arises from the work of any one trade. For instance, the way in which windows are set in a window opening determines the width of the window sills. There is no standard way of doing this, and if the drawings define exactly the way in which windows are to be set there will be no confusion. If they are wrongly fixed it is clear that it is the bricklayer's responsibility to take them out and build them in again as per the drawing. If it was not clearly shown in the drawing the issue would not be easily resolved. There are literally hundreds of such details in every construction job.

Construction drawings are also used by sub-contractors to design their services. An electrician will require a drawing which he will mark up with the wiring layout, and the plumbing and heating engineer will want them for the same purpose. Others will be required when the kitchen is being planned. Central heating drawings are often provided free of charge by the fuel advisory agencies, and the kitchen layouts can be obtained from various bodies, but all start with a print of the actual construction drawing.

The drawings will have been drawn to either metric or imperial scale, although both dimensions may be shown. These days imperial design work is unusual, and may be identified by the scale, typically ¼" to 1' or ⅛" to 1'. Metric scales are typically 1/100 and 1/50. All setting out of construction work should be done in the units used for the design, and the converted dimensions used with considerable care as they are invariably 'rounded off' and if added together will give rise to significant errors.

Remember that room sizes on construction drawings are masonry sizes and that the finished dimensions from plaster surface to plaster surface will be about 25 mm (1") smaller. Carpet sizes will be the same amount smaller still, allowing for skirting thicknesses.

A full set of drawings for a traditionally built home are reproduced overleaf, and are miniatures of metric A1 size drawings. Drawings with this amount of detail are only normally obtained from the specialist 'package deal' firms. Some of the information on the drawings may be superfluous for the experienced tradesmen, but much is essential.

Revisions to drawings are normally made by altering the master drawing. When this is done the fact that the drawing has been altered should always be noted on it, and dated. Prints of the unrevised drawing which are now out of date should be carefully collected and destroyed to avoid confusion

For complex projects drawings are normally accompanied by a specification (written by an architect), and a bill of quantities (compiled by a quantity surveyor). Between them these highly technical documents describe and define every detail of the building and are only relevant to a building contract that is being supervised by an architect. However, a simple specification, usually referred to as a short form of specification, can be of use in placing a contract with a builder. We reproduce a short form of specification on pages 50 and 51. This specification covers all aspects of the construction in general terms, and would normally be adequate to describe the work detailed in a contract.

The drawings in this series are reduced to less than a quarter of their original size to facilitate reproduction in this book. As such they are barely legible, but they are useful to show the sort of drawings which are required for a house built with simple foundations on a level site.

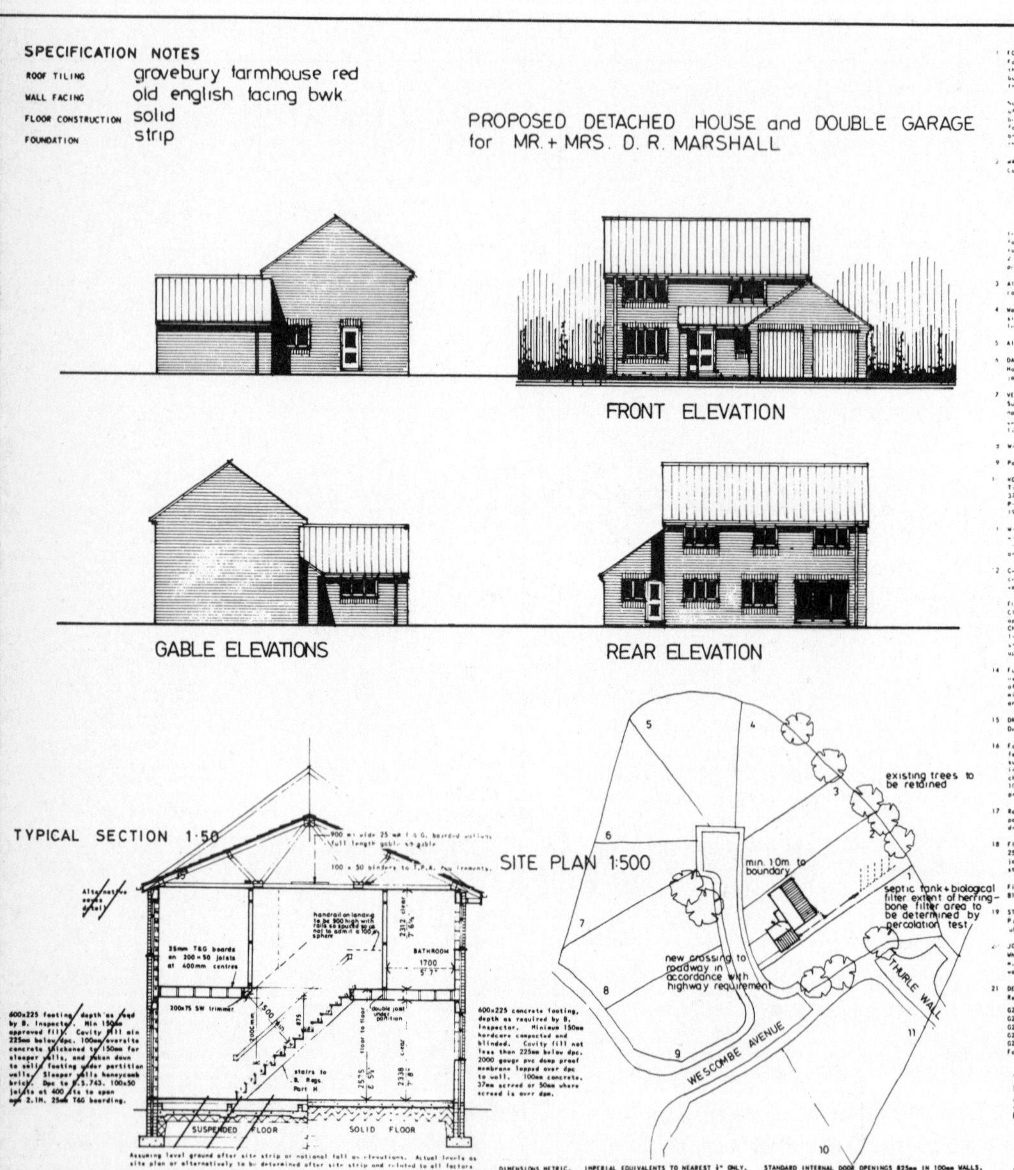

This drawing shows the four elevations of the house at a scale of 1:100 with the description of both the brick and the tiles to be used. It also includes a 1:500 site plan, essential in order to enable it to be used to support an application for planning consent.

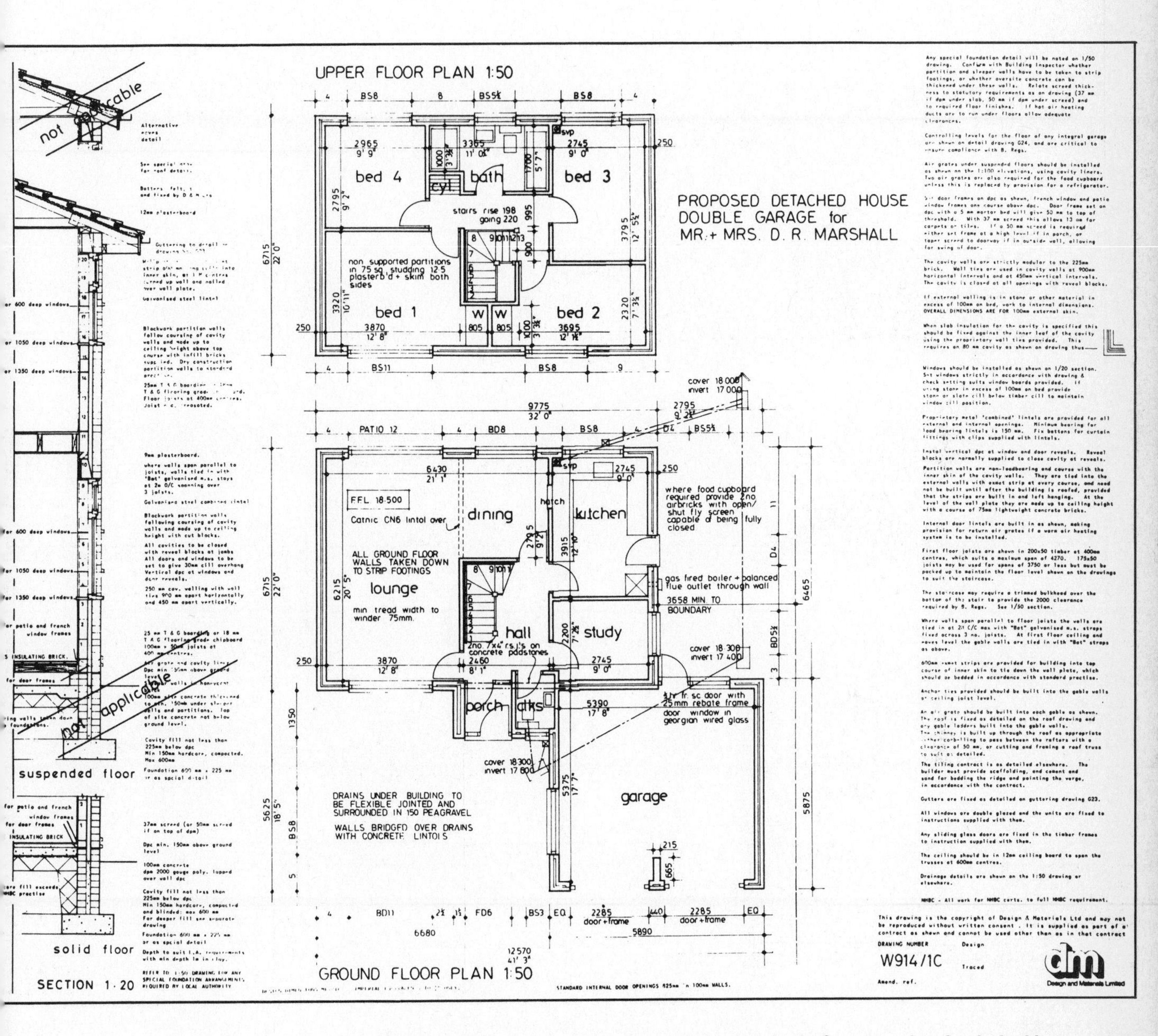

This 1:50 plan will also have to go to the local authority for approval under the building regulations. The data in the side panels define all the key construction detail, and also confirm that many aspects of the construction will conform with the building regulations.

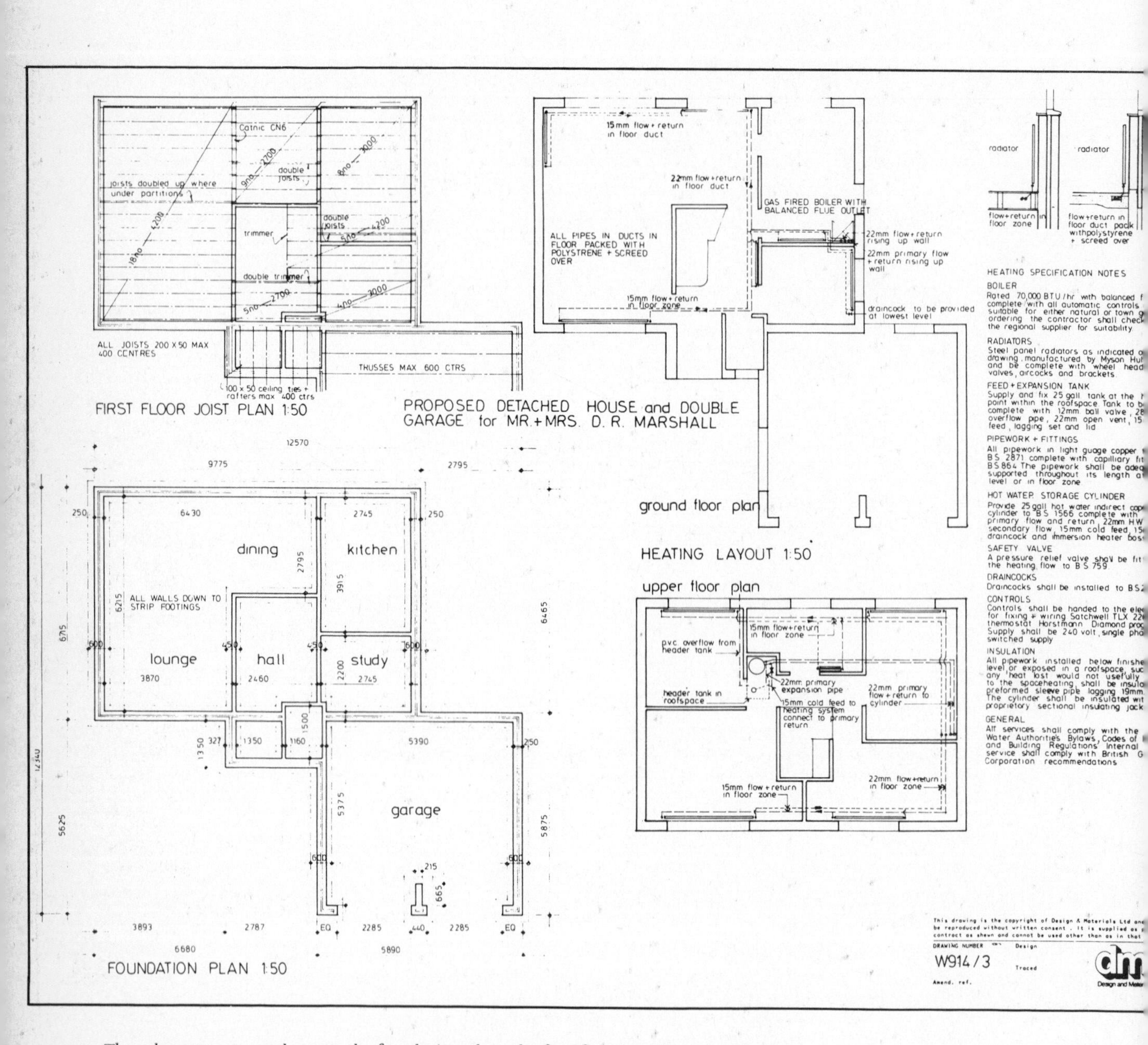

Three layouts on one sheet — the foundation plan, the first floor joist layout, and the heating arrangements with all the data and specifications for the central heating installation. Spare copies of this drawing can be over-drawn to plan the electrical layout, the kitchen layout, etc..

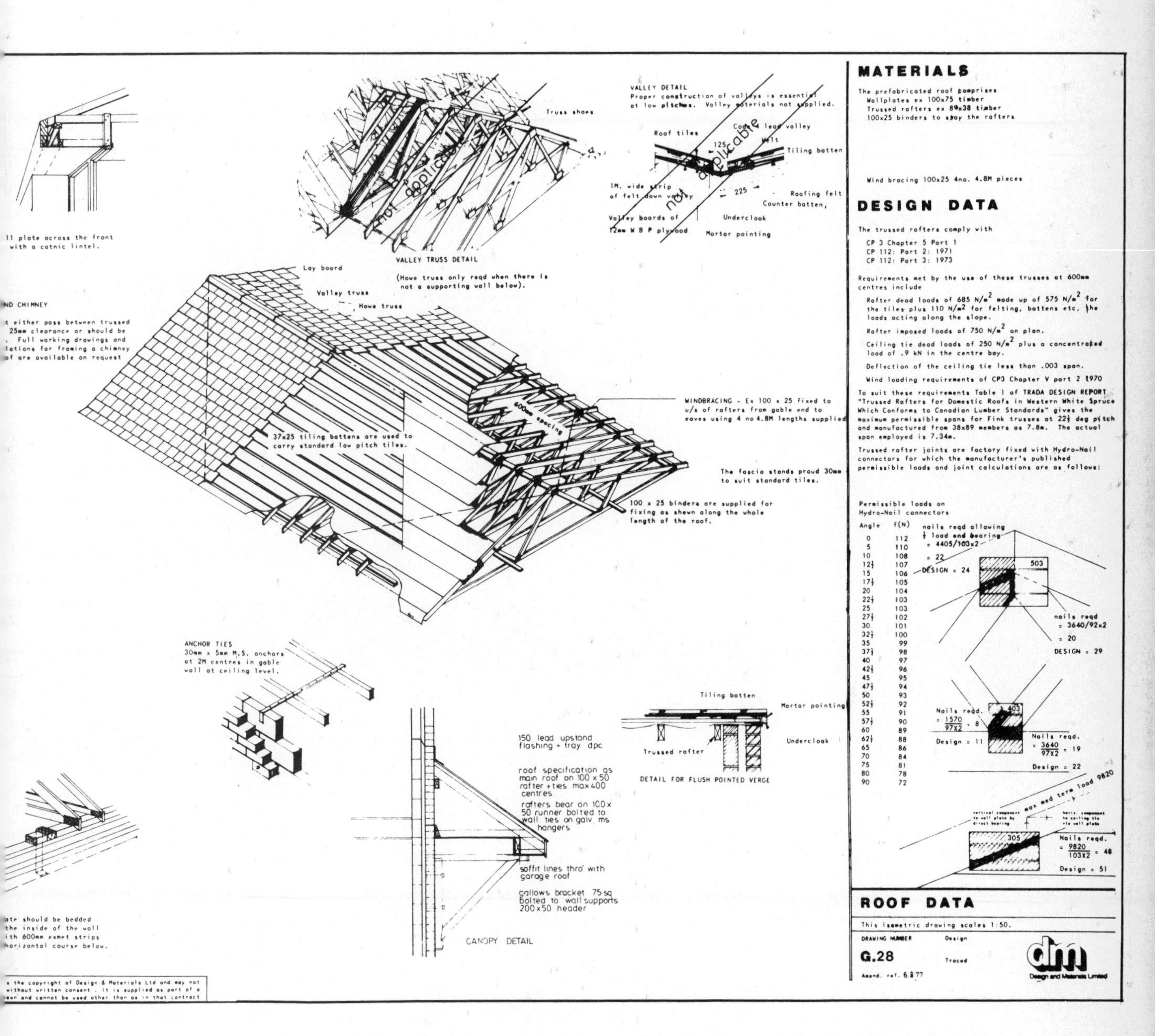

The roof construction drawing, with design data and details of the porch roof.

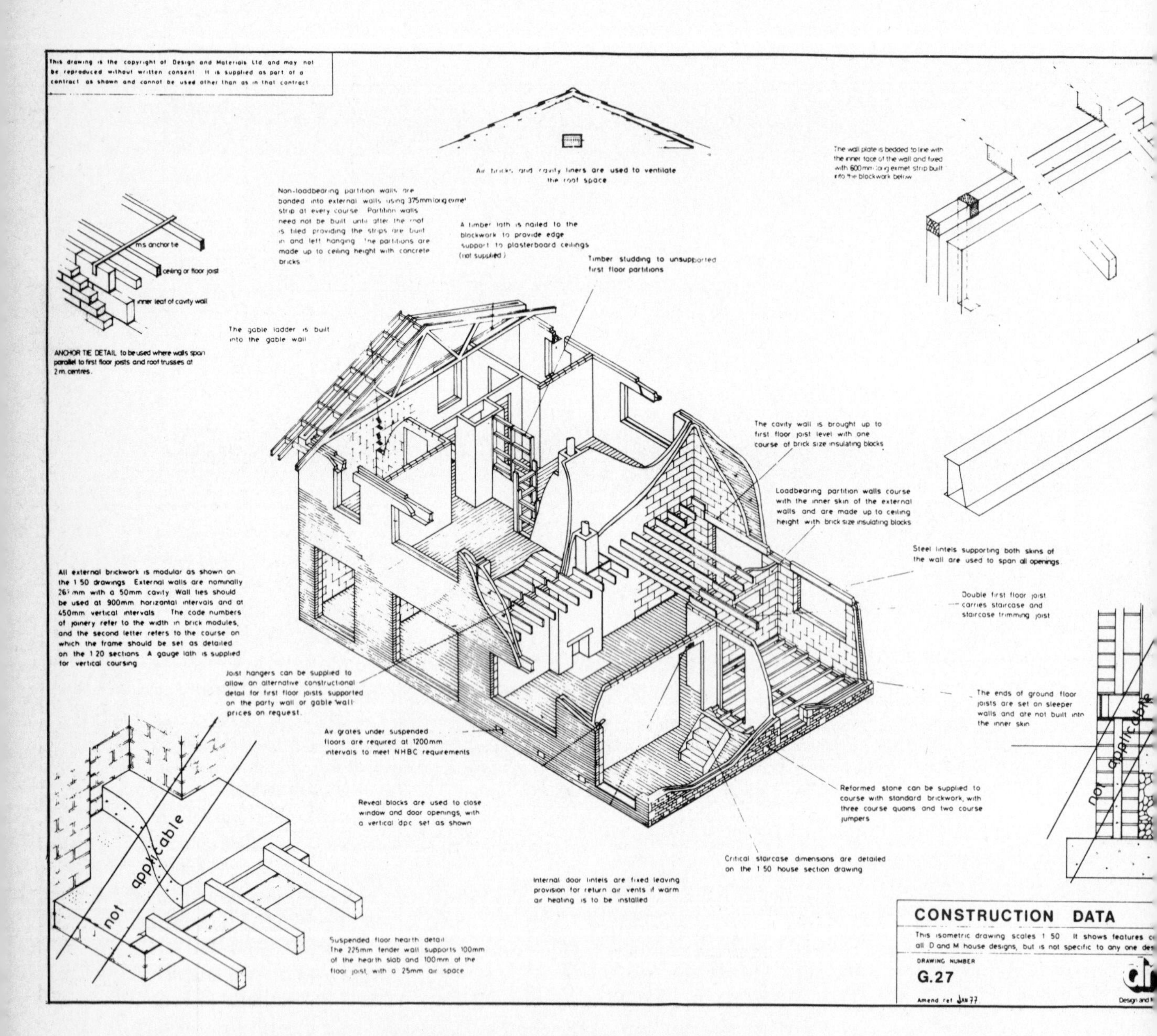

This isometric construction drawing gives a great deal of additional detail in a particularly graphic way, and helps to avoid mistakes. This is a standard drawing, and some of the information is not relevant and has been marked accordingly. Drawings of this type are normally only supplied by the national companies who supply the materials. Modern computer controlled tracing machines can produce drawings like this automatically once data from plans and elevations have been digested by their microchips.

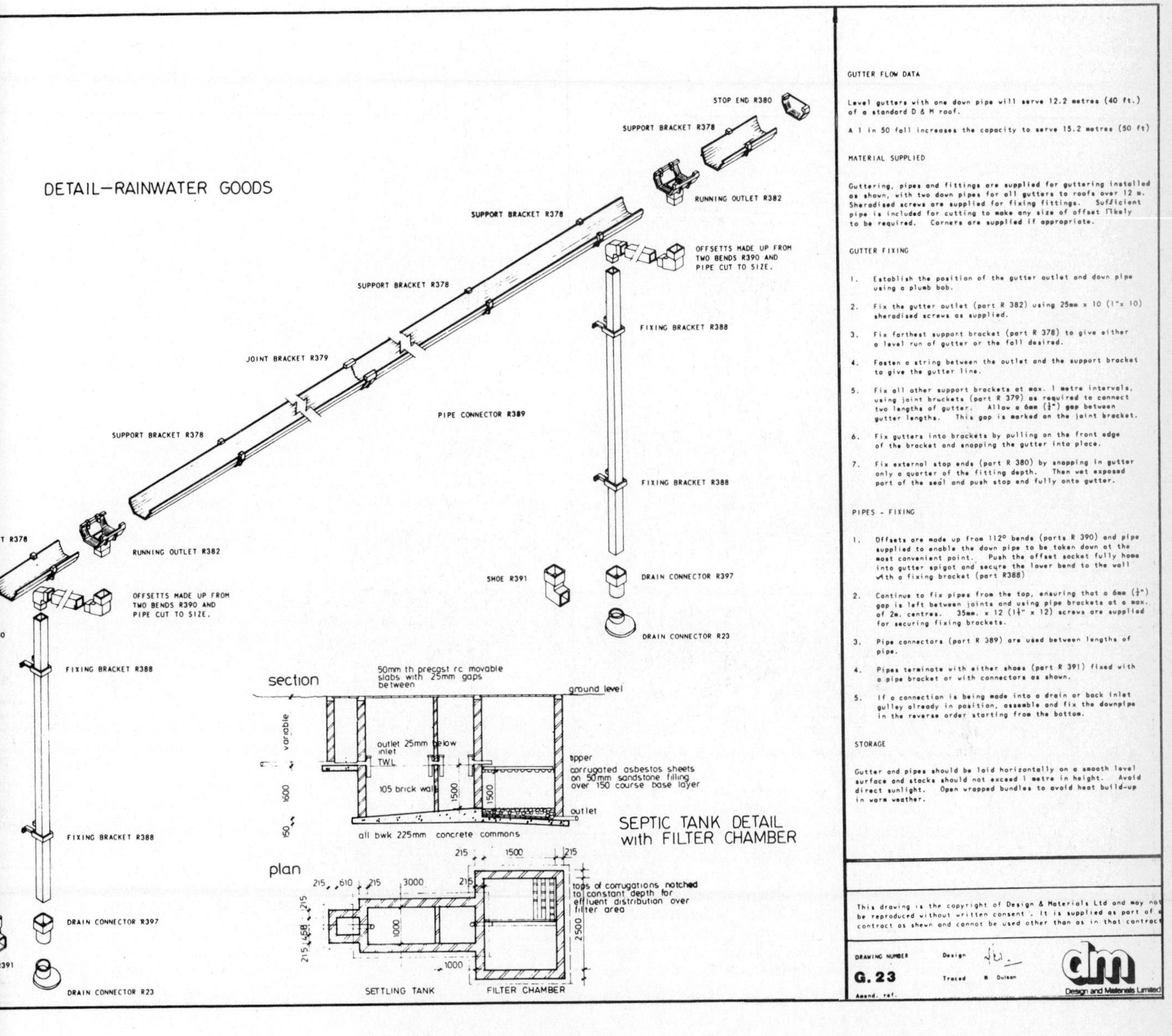

An isometric drawing showing the components and assembly of the gutter system, together with a plan and elevation for the septic tank. This latter detail has been tucked away on this drawing in order to reduce the total number of plans for this particular building.

TYPICAL SHORT FORM OF SPECIFICATION

A full architect's or quantity surveyor's specification is a long and complex document, couched in technical jargon, and defining materials, methods of construction and standards. Relatively few new homes on clients' own land are built with full specifications, and many are built without the essential definitions of key elements in the builder's contract. This short form of specification covers these definitions, but avoids technicalities by referring to well known published standards, particularly the NHBC Handbook.

This specification is an example only, and is intended as a guide to drawing up a specification for a specific contract for a specific house.

1. This specification relates to a contract established by . . . (detail form of contract, if any, or exchange of quotation and acceptance which establishes the contract) . . . and is a schedule to that contract. The contract is between the client . . . (name and address) . . . and the builder . . . (name and address) . . . and is a simple contract between the client and the builder. Neither the designer, any supplier of materials or any sub-contractor are party to this contract.

2. This specification refers to a house to be built to the drawings attached, which have been initialled and dated by both parties, and all notes on these drawings are part of the specification.

3. The builder shall obtain an NHBC certificate for the property, and shall provide the client with the documentation relating to this in accordance with standard NHBC practice before work commences. All materials and work shall be to the requirements of the NHBC Builders Handbook, and shall follow the further recommendations laid down in the NHBC site manuals and practice notes.

4. Time will be the essence of the contract and the builder is to start the works on . . . (date) . . . or as soon as practicable thereafter, and shall finish the whole of the works in the time stated in the tender.

5. Two sets of working drawings will be furnished to the builder for site use, and any further prints reasonably required shall be supplied on request.

6. The term prime cost when applied to materials or goods to be fixed by the builder shall mean the list price of such goods as published in the supplier's catalogue, and any trade discounts obtained by the builder shall be an advantage enjoyed by the builder. Prime cost sums shall include suppliers' charges for delivery. All expenses in connection with the fixing of such goods shall be allowed for by the builder in the contract sum.

7. All work and materials shall be to British standards and Codes of Practice and shall comply with Building Regulations. Proprietary materials and components shall be used or fixed in accordance with the manufacturers' recommendations.

8. The builder shall be responsible for the issue of all statutory notices and shall comply with the requirements of the Local Authority and statutory bodies. The client warrants to the builder that all necessary planning consents and appropriate building regulation approvals have been obtained, and shall be responsible to the builder for any delay or cancellation of the contract consequent on there not being such consents or approvals.

9. The builder shall be deemed to have visited the site and to have satisfied himself regarding site conditions.

10. The builder shall be responsible for all insurances against all risks on site, including public liability and fire risk, to date of hand over. The builder shall make security arrangements for the proper storage of materials on the site as appropriate to the local circumstances. The builder is to avoid damage to public and private property adjacent to the site, and to make good or pay for reinstatement of any damage caused. The builder shall extend to the client the guarantees available to him on proprietory materials and fittings, and shall provide the client with documentation required to take advantage of such guarantees.

11. The builder is to cover up and protect the works from the weather, and to take all action for the protection of the works against frost in accordance with the requirements of the NHBC.

12. Top soil shall be stripped from the site before commencing excavation of the foundations in accordance with the requirements of the NHBC, and shall be . . . (spread or left heaped) . . . Any trees removed shall have the whole of their roots excavated, and the back fill shall be with material appropriate to the works to be executed over the excavation.

13. The builder is to set out and level the works and will be responsible for the accuracy of the same.

14. Foundations shall be as per the drawings with footings under partition walls taken down to solid ground. Depth of the foundations shall be as per drawings, with any additional depths required by the Local Authority paid for at measured work rates.

15. Concrete for the foundations and solid floors shall be truck mixed concrete as specified. Foundation brickwork shall be in bricks or blocks to the requirements of the Local Authority. Fill shall be clean material to the requirements of the NHBC.

16. Ducting shall be provided for service pipes and cables through the foundations, and chases shall be formed in concrete for pipework inside the building in accordance with good building practice.

17. The ground floor if to be of solid construction shall have a sand cement screed, finished to receive . . . (tiles as defined or carpeting) . . . or if suspended floor construction shall be as per drawing with floor boarding to be . . . (define whether tongue and grooved boarding or interlocking flooring panels) . . .

18. Mat wells shall be provided at the front and back doors to be . . . (define type and size) . . .

19. The shell of the building is to be built with the materials specified on the drawings. The external walling material shall be . . . (make and type) . . . and shall be laid and finished in a manner to be agreed. The windows shall be . . . (make and range) . . . and shall be finished . . . (define finish). The external doors shall be . . . (make and types) . . . and shall be finished . . . (define finish). Other external joinery, including any cladding, fascias and barge boards shall be finished . . . (define finish). Internal door frames shall be . . . (material and finish). Window boards shall be . . . (material and finish) . . . Staircases shall be . . . (material and style, particularly style of balustrade and rails).

20. The roof shall be constructed strictly in accordance with the drawings, and any trussed rafters specified shall be of a type for which building regulation approval has been obtained.

21. The roof and any vertical external tiling shall be tiled with . . . (define tiles by manufacturer, type and colour) . . . and the tiling work shall be carried out by a tiling contractor approved by the tile manufacturer so as to obtain the most favourable guarantee available from the tile manufacturer. The subcontract shall be between the builder and the sub-contractor.

22. First floor boarding shall be . . . (define whether tongue and grooved boarding or interlocking flooring board) . . .

23. Access to the roof shall be provided to NHBC requirements, and a loft ladder shall be fitted within the contract sum.

24. All walls shall be plastered in . . . (define lightweight or traditional plaster) . . . to the plaster manufacturer's full specification, and all materials used shall be from the same manufacturer. Coveing and other plaster features shall be extras to the contract, to be specifically defined in a quotation and ordered with a written order.

25. Ceilings shall be boarded to suit the ceiling finish specified, which shall be . . . (define) . . .

26. The under surface of the stairs shall be . . . (define arrangements for below the stairs if this is a visible feature) . . .

27. Architraves and skirtings shall be . . . (define material, size and moulding shape after discussion of samples) . . .

28. Internal doors to be . . . (define doors specifically, by manufacturer and model) . . .

29. Any sliding patio doors shall be . . . (manufacturer and type) . . .

30. All windows shall be double glazed with sealed double glazing units to be . . . (manufacturer and type) . . . All glazed doors shall be single glazed. Obscure glass shall be used for glazing to . . . (define rooms) . . .

31. Garage doors shall be . . . (manufacturer and type) . . .

32. The door furniture to be as the schedule attached (The schedule should detail which internal doors are to have latches, which are to have locks, security locks to external doors, letter plates as required, plus any other fittings. Windows are supplied complete with furniture, but if security bolts or special fittings are required these should be specified.).

33. The central heating system shall be installed against a prime cost sum of £ . . . and the proposals for this system shall be as detailed separately. The system shall be designed to meet the heating requirements of the NHBC, and all work shall be to the appliance manufacturer's requirements. If the heating system requires the installation of an oil tank, the position and height of this shall be agreed, and the structure to support and/or conceal the tank shall be . . . (define) . . .

34. The chimney and chimney breast shall be built as per drawing, and the fireplace opening provided shall be for a . . . (name appliance) . . . This appliance and the fire surround shall be provided against a prime cost sum of £ . . . All work to the fireplace opening and chimney shall be to the appliance manufacturer's requirements.

35. Sanitary ware and bathroom fittings shall be provided against a prime cost sum of £ . . . (discuss).

36. The cold water tank shall have capacity of . . . (discuss) . . . and shall be situated in the roof in a position agreed, to give ease of access, on a stand to NHBC requirements. It shall be fitted with a lid, and frost protected as required under the Building Regulations. The hot water cylinder shall have a capacity of . . . (discuss) . . . in a position to give ease of access while providing for the maximum space for shelving alongside. The hot water cylinder shall be fitted with an immersion heater, to be . . . (discuss, including whether this is a dual model to provide both full and top-up heating). The cylinder shall be lagged to NHBC requirements.

37. The kitchen fittings shall be provided against a prime cost sum of £ . . . and this shall include all sinks down on drawings. Hot and cold water and drainage connections to a washing machine/dish washer situated . . . (define) . . . shall be provided.

38. Wardrobes, cupboard fronts and other fitted furniture shall be provided against a prime cost sum of £ . . .

39. The electrical installation shall allow for lighting points, power points and switching arrangements to be to the NHBC minimum requirements, and the builder shall quote the additional sum required for each extra ceiling light, extra wall light, and each extra socket outlet. Light switches and socket outlets shall be . . . (manufacturer and range) . . . Simple pendents shall be provided at all lighting points, or alternatively the client's fittings will be fixed if provided to programme. The fuse box shall be . . . (define fuse board and circuit breaker system) . . . Provision shall be made for television sockets and telephone points in . . . (define rooms and position in rooms) . . . Electricity meters shall be in a meter box fitted . . . (define position. The Electricity Board may try to define where this should be) . . . Provision shall be made for bells at the front and back door, and a simple bell shall be provided, or alternatively the client's chimes or other fittings to be installed if provided to programme.

40. All interior plaster surfaces shall be finished with . . . (define emulsion paint and colour) . . . which shall be applied in accordance with the manufacturer's recommendations for new work to give a consistent colour.

41. Wall tiling shall be quoted as a prime cost sum of £ . . . per sq yd. for a stated minimum area.

42. All Softwood joinery shall be knotted, primed and treated with two undercoats and one gloss finishing coat of interior paint to be . . . (define paint and colour) . . .

43. All timber surfaces which are not to be painted shall be protected by using Sadolin or similar protective stain to the manufacturer's requirements, and shall not be varnished (or other requirements as considered appropriate).

44. Foul drainage shall be as detailed on the site plan, and all work shall be to the requirements of the Local Authority.

45. Rainwater goods shall be . . . (manufacturer and range) . . . and shall discharge into open gullies or via sealed drain connectors . . . (as defined) . . . Surface water drains shall discharge into soakaways or elsewhere as detailed on drawings.

46. External steps, and the porch or step at the front door, shall be finished with . . . (quarry tiles or finish required) . . . A path shall be provided around the whole of the perimeter of the building, to a width of 2 ft., to be . . . (specify concrete surface or paving slabs) . . . and shall be laid to the full requirements of the NHBC for external works.

47. Other external works, including any work on the drive, or any fencing, shall be considered as extras to the contract, and shall be quoted for in writing, and the order for them placed in writing.

48. Any detached garage shown on the drawings is outside the contract, and any work to construct such a garage shall be an extra to the contract, to be quoted in writing and any order placed in writing.

49. Any other work required or fittings to be installed in connection with the contract shall be quoted in writing and any order placed in writing.

50. Any defect, excessive shrinkages or other faults which appear within 3 months of handover due to materials or workmanship not in accordance with the contract, or frost occurring before practical completion, shall be made good by the builder, and payment of the retention detailed in the payment arrangements at paragraph 51 shall only be made on completion of this making good.

51. Payment to be made on a progress rate of:

20% of contract price at dpc
25% of contract price at roof tiled
25% of contract price at plastered out
30% at handover

The above all subject to a 2½% retention as provided in paragraph 50. All payments to be made within 7 days of notice that payment is due.

52. The client may but not unreasonably or vexatiously by notice by registered post or recorded delivery to the builder forthwith determine the employment of the builder if the builder shall make default in any one or more of the following respects:

i) If the builder without reasonable cause fails to proceed diligently with the Works or wholly suspends the carrying out of the Works before completion.

ii) If the builder becomes bankrupt or makes any composition or arrangement with his creditors or has a winding up order made or a resolution for voluntary winding up passed or a Receiver or Manager of his business is appointed or possession is taken by or on behalf of any creditor of any property the subject of a Charge. Provided always that the right of determination shall be without prejudice to any other rights or remedies that the client may possess.

53. The builder may but not unreasonably or vexatiously by notice by registered post or recorded delivery to the client forthwith determine the employment of the builder if the client shall make default in any one or more of the following respects that is to say:

i) if the client fails to make any interim payment due within 14 days of such payment being due.

ii) if the client or any person for whom he is responsible interferes or obstructs the carrying out of the Works.

iii) if the client becomes bankrupt or makes a composition or arrangement with his creditors.

Provided always that the right of determination shall be without prejudice to any other rights or remedies which the builder may possess.

54. In the event of a dispute between the parties arising out of the contract, the parties shall agree jointly to engage an architect independent of either of them to arbitrate between them, and shall be bound by the architect's findings as to the matter in dispute and to his apportionment of his fees as an arbitrator.

Tackling the Job

It is very unusual for a self-builder to put up a new home for himself without some relevant experience. This might be in the building industry, giving a familiarity with its structure and procedures if not of one particular trade, or else in management, with a knack for getting things done. One large category of individual self-builders are farmers and small-holders, and certainly they are the most successful group. Their strength lies in their versatility, since dealing with new situations and acquiring new skills is common-place. They are used to handling money, can be philosophical over the risks involved, and are familar with dealing with both officialdom and sub-contractors. Self-employed artisans and businessmen have the same attributes, particularly if they work in the building industry. Self-builders directly employed on a regular basis are fewer in number outside housing associations, and salary earners as opposed to wage earners are fewer still. What all self-builders have in common is enormous self-confidence, backed with experience of how to get things done. Knowledge of building techniques is certainly not the essential ingredient of success, while a knack for finding the right answer to every problem definitely is.

The success of a project is usually determined by the project management before any work starts on site. By the time the technical—legal—financial infrastructure has been established you may be forgiven for thinking that now surely is the time for simple skill and hard work. After months of managing paper and people, formalities and delays you will look forward with pleasure to the actual building work—hard, unremitting toil but very worthwhile—the creative part of the whole exercise.

This is fine since you will want all the enthusiasm you can find, but it is essential that the work is managed as carefully as all the earlier stages. As stated so often before, by building on your own you are working outside the system, and the only way to do it successfully is by planning every move—including the donkey-work. 'Get stuck in' is a useful motto only as long as it is to a carefully planned construction programme based on a careful analysis of the job. The phases of the construction work involving different trades have to be programmed to follow in sequence, and deliveries of materials and the provision of site services have to be arranged to suit. Nothing will destroy the enthusiasm for work on site more quickly than confusion or delays, while the best boost to morale is the steady progress of work to a realistic timetable. Management is 90% of the building operation.

For those in housing associations who have adopted the federation's model rules this is all set out for them, with officers of the association appointed to deal with every aspect of the site management. The officers—secretary, foreman, time-keeper, purchasing officer, and the rest—must work as a committee and realize that not only have they to act as a team, but that their primary function is to maintain the morale of the members and the impetus. Planning the work, and arranging deliveries and services on schedule, ensure that this happens. Given this, the earth will be moved, the bricks laid, and the roof topped out. The Wadsworth self build organisation has the slogan 'Given the right organisation anyone can build his own home'. Hundreds of successful self-build schemes since 1945 have proved it to be true and schemes that fail have the roots of the failure in a lack of organisation.

With an effective committee the self-build group has every advantage over the individual. They can allocate trades between them, with experienced tradesmen training the amateurs, lending their professionalism to the project. Jobs like unloading or laying concrete are more easily tackled by a team, and above all members lend each other the moral support necessary to make 1000 hours of part-time work a year practicable.

The individual, lacking these advantages, needs to be cast in an even more heroic mould than the most enthusiastic group member. Without a committee to plan ahead, he is more liable to attempt to muddle through, and to take on more than he should. He must plan the job, and if he is contributing his own labour, he must decide how many hours he can put in, and build this committment into his programme. He has to analyse what he can do, what he can learn to do, and what he should do. These are unlikely to be the same. Anyone can learn to mix concrete. Given a set timetable and a fixed number of hours of personal involvement it will probably be more cost effective, although more expensive, to buy truck mix concrete. Anyone can dig trenches, but a JCB will dig them at a speed that will make the hire charge more cost effective. Almost anyone can learn to bone a line along a trench and lay drain pipes, and if he applies his labour here the savings will be really significant. Amateur roof tiling is hazardous, back-breaking and unlikely to be worthwhile when a supply and fix tiling contractor costs little more than the cost of the tiles alone. On the other hand, sub-contract labour rates for brick-laying are now often over 10p per brick laid, making this the most desirable of skills.

All decisions about the jobs tackled, and those put out to contract, depend on individual circumstances, but there are three areas where the self-builder should be extremely cautious— electrical work, gas fitting and plastering. The hazards of the first two are self-evident, and in spite of the checks made by the gas and electricty boards before connecting main services, it is morally indefensible to run the risk of making installations with a built-in hazard. Plastering work is not only a particularly skilled trade, but a sub-standard job cannot be remedied or disguised. The author has never seen a first class

Members of the Southfields Self Build Housing Association mixing concrete. In most circumstances this is not cost effective and it is invariably better to buy truck mix concrete.

amateur plastering job, but has met many where an uneven wall or rippled ceiling is a monument to over-enthusiasm.

Virtually every other skill required in building a house can be acquired by a competent enthusiast who sets out to learn them methodically. He will be exceptional, but hundreds like him build their own houses, and facilities for learning basic trades are provided at evening classes and adult education centres. Usually a self-builder will develop one particularly good skill, for which he has a natural flair, and make it his business to find out enough about the others to be able to place sub-contracts and to supervise them. Besides the courses organised for those who wish to acquire building skills there are innumerable books and trade magazines available. Among the latter the Building Trades Journal is a must for self-builders, and will give a useful insight into materials, prices, labour rates and many technical matters.

Site safety is another field where prudence and management must take precedence over enthusiasm. Building is a hazardous occupation and the amateur will be more at risk than the professional. The most spectacular construction site accidents involve contractor's plant, scaffolding or deep trenches, but the great bulk of them stem from trivial mishaps. Typical incidents come from working off improvised scaffolding, however low, or from walking into protruding scaffold putlocks, or from waste materials negligently dropped from a height. While the author was visiting one Housing Association site to collect material for an earlier edition of this book a member trod on a discarded floor board left with a protruding nail. The nail went through his foot, depriving the association of a team member for weeks. Another hazard is that of cement burns from concrete or mortar finding its way down wellingtons and being allowed to stay there. As the seventh edition of this book went to print a Cambridge self builder was in hospital undergoing skin grafts to repair injuries caused this way. Most building manuals have lengthy sections on site safety; in practice 99% of accidents can be avoided by running a tidy site and by using common sense. Most self-builders are either obsessively tidy or unbelievably untidy, and it is worthwhile being the former.

There are various statutory requirements for those who run building sites, Provision of latrine facilities, a hut for meals, protective clothing, a first aid box and accident register are required by the Factories Inspector who is unlikely to visit the site, but most self-builders will provide these as a matter of course. Neglecting these regulations could have serious consequences for those responsible, so do not keep the petrol for the mixer in the mess hut and use the top of the tin as an ash tray, or allow children to ride on a dump truck. In law a self-build building site is just a building site like any other, and the rules are strict. Most associations have members employed in the building industry who know about this, and one of them should be appointed safety officer under the model rules. The individual self-builder can only do as he thinks fit in the light of the problems that would arise if he killed himself.

Unlike the Factories Inspector, the Building Inspector certainly will be seen on site, and it is a condition of building regulations that he is given due notice of the progress of work so that he can make his statutory inspections. Local authorities supply post cards for this purpose and these should always be sent off in good time. As discussed earlier, it is essential that self-builders should regard the Building Inspector as a friend and adviser, and avoid the 'him and us' attitude which some sections of the building industry adopt. Remember that he has a statutory duty to make these routine inspections, and if work is covered up before he has checked it, he can require it to be exposed or taken down. He is a good friend, or a very bad enemy, and if you stick to the rules he will invariably be the former.

Another visitor will be the architect, surveyor or value who may be appointed to satisfy the bank or building society that all is in order, and to issue routine progress certificates. These gentlemen are as difficult to satisfy as Building Inspectors and not nearly as prompt in getting out to the site when requested. However, as their visits result in the release of funds these disadvantages are happily accepted. Remember—if in doubt, give any official visitor a mug of tea. Don't be ashamed of the short-comings of either the mug or the tea; if he has been any time in the game he will have seen and tasted much worse.

Purchasing

Purchasing materials and arranging deliveries are matters where a little experience is worth a great deal of theory. Here again, associations are able to appoint purchasing officers who have building industry experience, while the individual is on his own. Brief notes on buying procedures for different materials follow, and beware special arrangements of any sort without being sure what is involved. There are few bargains to be had in basic materials, but fittings and fixtures can often be bought at a huge discount. However one buys, the best terms are always given to those who pay promptly, and who are buying as trade purchasers. If in doubt ask to see the manager, tell him what you are doing, that you can pay cash, and ask for a special discount. You may get an extra 2½%!

Sources of Supply

An account with a builder's merchant is essential to any building operation for the supply of the hundred and one small items required such as wall ties, lintels, concrete additives, nails, timber preservatives, as well as the cement, sand, and other 'heavy' materials. Large builder's merchants can supply all the materials for a new house, but most self-builders buy materials as follows:

Bricks: Either direct or through merchants. Prices are standardised. Common bricks more often purchased through merchants, facing bricks more often direct. Can be mechanically off-loaded at an extra charge.

Blocks: Usually through merchants. Can be mechanically off-loaded at extra charge.

Joinery: Direct from manufacturers.

Roofing trusses: Direct from manufacturers.

Timber: From timber merchants who may also be roof truss manufacturers.

Tiling: Tiles from manufacturers, or supply and fix tiling from tiling contractors.

Glass: Glazing contractor or double glazing manufacturer.

Plumbing materials: Plumber's or builder's merchants, often using advantageous special offers on bathroom suites and central heating equipment.

Kitchen fittings: Wherever the best offers are available.

Roof insulation: Best available special offers.

Plant and scaffolding: Individuals hire from specialist hirers, associations buy second hand and re-sell after use.

Everything else: Builder's merchants. Some large builder's merchants have trade 'cash and carry' arrangements, which present accounting complications but offer very low prices.

Prices are only half the story in buying building materials. Delivery, when and as required, is equally important, and suppliers who can deliver at the unusual times required to suit self-builders may be the ones to look out for. Suppliers require customers to provide labour to unload.

Ordering procedures in the building industry are casual by most standards, and it is not unusual for hundreds of pounds worth of materials to be delivered to established customers on the strength of a phone call, or even a verbal message given to the supplier's lorry driver. If an association has a purchasing officer who is part of the building industry he will know how to operate like this but for the newcomer written orders are to be recommended.

All building materials that are not fixed by the supplier carry VAT, and to reclaim this every invoice must be kept. However casual the accounting procedures, from the meticulous book-keeping required of associations to the informal accounts kept by many individual self-builders, VAT invoices must be kept and used to claim back the tax charged.

Sub-contractors and Skilled Labour

Employing skilled or unskilled labour for the building work is relatively easy, but it is important to be clear exactly what arrangements are being made. There are different ways in which you may arrange for the work to be done. Certain jobs are invariably on a 'supply and fix' basis, with the sub-contractor supplying all his materials. Sub-contractors will quote from your drawings and should define clearly what they propose to supply, and what you have to supply. Orders should always be placed on the basis of a detailed quotation, and should cover everything discussed with the sub-contractor.

Sub-contractors who supply both labour and materials may be able to offer a Building Employers Confederation guarantee. This is a new scheme, introduced in late 1984, and is likely to have a significant effect on various aspects of the self-build movement. The BEC was better known as the Federation of Building Trades Employers until it changed its name, and is one of the most influential trade associations in the building industry. The new guarantee scheme is primarily designed to re-assure those who employ its members to handle repairs and alterations, but it is appropriate to the sort of work done by sub-contractors for self builders. The guarantee is against bad workmanship, or a contractor going bankrupt, or defaulting. This scheme may become as universal as the NHBC certificate for new homes, and if so, then perhaps BEC guarantees on sub-contractors work *may* become a condition of bank or building society finance. An interesting thought; at this stage it is in its early days, but is worth watching.

The BEC scheme is administered by the BEC Building Trust at Invicta House, London Road, Maidstone, ME16 8JH, and they will supply details of how it all works and of BEC members in any particular area.

Looking at 'labour and materials' subcontractors first, the different tradesmen with which you will be involved are:

'Labour and materials' subcontractors

Tilers

Tilers will felt, batten and tile the roof and will supply all their materials. They will also do any vertical tile hanging required. You supply access to the roof (i.e. leave the scaffold up until they finish), cement and sand for pointing (but they supply colouring to match the tiles) and arrange for the plumber to work with them where any flashings or leaded valleys are involved. You also provide labour to unload the tiles and stack them near the building. All carpenter's work on the roof should be completed before they start, and valleys to be tiled with valley tiles must be boarded. Ensure that the quote is for 'tiles fixed to manufacturer's recommendations', and send for your own copy of the recommendations. Check that tiles are nailed or clipped to suit your particular exposure rating—again see the manufacturer's recommendation. Establish the guarantee offered for both the tiles and for the tiling work.

Plumber

The plumber will normally quote for all materials, including bathroom suites and the sinks to install in your kitchen units. In his quote he is allowing for the benefit to him of the discount which he can get in buying these fittings on his trade account, so if you intend to buy them yourself you should make this quite clear. Remember—if he supplies the fittings then he is responsible for breakages until he is finished. If his quote includes a prime cost sum for fittings of your choice, which he will obtain for you, this means that you will have a saving on his quotation if you spend less than this total, and an extra if you exceed it. Always get him to define whether the p.c. sum is for the retail price of fittings, or for some other price, typically retail less 10%. Incidentally, the plumber will get about 30% discount, but he allows for this in his quotation. Always define the smaller fittings as well as the major ones, particularly taps. Establish whether tank and cylinder lagging is included in the quotation. Clear who has responsibility for arranging the water supply, and establish that the plumber has to arrange all liaison for water board inspections and testing.

A plumber does not usually give a guarantee as such, but should undertake to return to deal with any problems arising within a set period, typically six months. Finally, place an order on the basis that 'all work shall be to the standard set out in the NHBC Handbook'. You cannot get an NHBC certificate, but you can take advantage of their standards!

Heating Engineer

Your heating engineer may well be your plumber. Whether he is or not, there is the same position regarding the supply of appliances, and if you intend to buy either boiler or radiators yourself this should be made clear. Again, there are clear advantages in the sub-contractor providing everything, and being responsible for it until it is fixed.

The heating engineer must design the system to meet specified standards (such as those of the NHBC) or he should quote for a system designed by others, such as a fuel advisory agency. For the self-builder the latter is probably preferable. Ensure that the installer is responsible for commissioning the system, and if he does not have his own electrician, for 'proper liaison with the electrician having the site contract'. This avoids the situation where the installer announces that everything will start when your electrician has done his stuff, and that as far as he is concerned he has finished. Suitable guarantees on both equipment and installation should be offered.

Electrician

An electrician will normally quote a fixed price for an installation as shown on a drawing, or as detailed in a quotation plus a fixed extra charge for each additional light or power outlet required. He will supply the switches and sockets (of a make and type which should be specified) but will quote for simple batten or pendant light fittings only. As you will wish him to get the installation tested by the Electricity Board as soon as practicable, so that the mains connection can be made, he has to provide these fittings for the Board's tests. Normally you will be buying your own ornamental fittings, and if you can give them to him at the right time he will normally fix them free of charge in place of the pendants and battens. Do not ask for a rebate for the savings on pendants and battens—this is balanced by the cost of involvement with your own fittings, which you will have chosen for reasons having nothing to do with ease of installation!

The electrician will also fix T.V. points (but not T.V. aerials), telephone ducting and deal with your heating thermostat and appliance wiring. He will supply and fix any immersion heater required.

All electrical work should be offered as being to NHBC or equivalent standards, and it is important that you satisfy yourself that you are getting the best arrangement of ring mains etc. The fuse board should be accessible and have separate circuits clearly labelled. Discuss the possibility of the provision of contact breakers instead of fuses.

Plasterer

The plasterer will quote from your drawing for all plastering work in the building, including any external rendering, and for laying floor screeds. Breaking this down into the separate jobs:

Ceilings are boarded with plasterboard ready to receive a final surface of some sort. Note that if your ceiling joists are at 450mm centres you can use 9mm boards, otherwise you must ensure that 13mm boards are specified. Ceiling boards can be foil backed to provide added insulation if this is required. The finish is usually either 'skim' or 'Artex'. There are other similar materials, and they should be discussed with the plasterer. He will be able to arrange for you to see samples. If a proprietary ceiling finish is required, ensure that you specify that it shall be applied to the manufacturer's instructions.

Walls are either traditionally plastered, or dry lined. The former involves wet plaster being applied to the wall in two coats giving a solid feel to the resulting surface but taking up to two months to dry. The latter uses plasterboards on timber framing with a surface finish. Whoever designed the building will have specified walling materials to suit one of these plastering systems, but whichever is involved insist that work is done to the plaster manufacturer's specifications, and that if possible all the materials come from the same manufacturer. All modern plaster specifications include metal angle beading.

Plaster coving at the junction of a wall and the ceiling gives a prestige feel to any job, and is available in a number of styles. It is also relatively expensive. As an alternative, polystyrene coving can be installed by the decorator, although it does not give the same class of finish. If you intend to use polystyrene coving do not tell the plasterer!

External render can give a wide variety of finishes, and the costs involved also vary widely. The finish will almost certainly have been detailed in the planning consent, and should be applied to manufacturer's recommendations if it is a proprietary material, or in accordance with best local practice if it is a traditional finish. Aluminium render stop should be used to give a bell finish to the bottom of any rendered surface, and also to form a drip over windows and door openings.

A good *floor screed* can receive fitted carpets direct, without any other surface. The thermoplastic tiles laid in new speculative housing are there as a sales aid, and omitting them will save up to £3.50 per square metre. A first class screed is worthwhile, and usually the plasterer is the best man to lay it.

All plastering work should be quoted for on a lump sum basis, from the drawings, as the system of measurement employed for 'price per square metre' contracts is extremely complex.

All labour and materials sub-contractors can 'zero rate' VAT, which means that it is not charged at all, which is useful. Even more useful is the advice which these tradesmen will give to someone building on his own, and it is always worth asking them how they think the job should be done. Invariably they will be local people, who depend on their local reputation, and provided that normal commercial prudence is exercised in placing a contract, it is rare for difficulties to arise in connection with their work.

Insist on all work being carried out to material manufacturer's instructions. This is important because you will be able to refer any problem back to the manufacturer. Use of materials other than to manufacturer's recommendations will void their guarantee.

Labour Only Sub-Contractors

'Labour Only' sub-contractors are a mixed bunch, varying from those who are worthy descendants of the wandering medieval masons who were the original journeymen, to the worst of the 'lump' so reviled by the unions. The best introduction to a labour only sub-contractor is a personal recommendation from other tradesmen. It is worth asking anyone who you propose to entrust with your work to take you to see the last job he did, following which you can usually arrange to phone his last employer for a reference.

The labour only tradesmen likely to be met by the self-builder are:

Bricklayers

Most bricklayers come in gangs, usually called a '2 and 1' gang of two bricklayers and one labourer. They will have their own tools but will require a mixer, scaffolding, mortar boards, roller etc. They will lay bricks and blocks, and will usually pour concrete and lay drains. They should be able to set out the building and work to any drawing, and if they are local men who are

Note of an informal agreement made with a 2 and 1 bricklaying gang

John Smith to arrange for bricklaying work on bungalow drawing W719 as a self-employed sub-contractor.

All work as drawing and NHBC requirements and to satisfy Building Inspector.

Clean footings after JCB

Pour concrete for strip foundations.

Build up brickwork to dpc max 750 mm. Any extra depth at £xx per thousand bricks laid. Fill cavity as required.

Fill foundation with hardcore, consolidate, blind, lay membrane, etc. Pour slab.

Build shell to wall-plate. Bed wall-plate with assistance from carpenter.

Assist carpenter to rear trusses.

Build up gable walls and build in gable ladders.

Build up partition walls to ceiling joists.

Build up chimney through roof.

Lay drains as drawing, with I.C.'s, and connect to existing foul drains.

Lay surface water drains from gulleys to soakaways as drawing.

Lay 900 mm wide path on rolled hardcore all around building.

Unload all own material and tiles.

Employer to supply all materials, diesel mixer, carpenter to assist with wall-plate and gable ladders, roller, water supply. Scaffold supplied to be Quickstage, which sub-contractor will erect and dismantle.

Work to start (date) and continue every practicable working day.

At lump sum price £xxx.

Payable £xxx brickwork to dpc, £xxx slab cast, £xxx building to square, balance when all work is completed.

settled householders, they will probably do it very well. Such gangs like to quote per thousand bricks laid, per metre of blocks laid, and per cube of concrete poured. Unless you are used to all this it is better to insist on a price for the job, to include all concrete work and drains that they are to be concerned with, with clearly defined stage payments. The only uncertainty should be any excess depth of foundation required, and this alone can be expressed as a price per thousand bricks. It is usual to get a verbal quote for such an arrangement, and it is a good idea to confirm this in a note with a copy for each party. A typical note is shown on page 59, and should prevent subsequent argument. The arrangement is made with one person only, the foreman of the gang, whose arrangements with his partners are his own affair.

Some bricklaying gangs specialise in work below dpc (damp proof course), are are known as groundworkers. Treat them like bricklayers.

Carpenters

Carpenters tend to be more settled than bricklayers, and often have a small workshop with limited machining facilities. A contract for the carpenter's work in a new house is in two stages, and treat these quite separately. The carpenter's 'first fix' is the roof, hanging the external doors, setting internal door casings, and fixing window boards and ceiling traps ready for the plasterer. The rest of his work comes later, when the building is plastered out. A carpenter will usually provide a written quotation.

714 Certificates

All labour only sub-contractors working within the building industry must be registered, and must supply a sub-contractor's registration certificate to their employer on engagement. If they do not produce this certificate the employer is obliged to remit 30% of the payment due to them direct to the Inland Revenue. Private employers outside the building industry are exempt from this. As a result, any advertisements for a sub-contractor to work on a private project will arouse interest from those who wish to avoid these regulations, and such interest needs to be evaluated with care. Registration certificates are often called '714 certificates' after the number of the form on which they are printed.

Casual Labour

Self builders who are already employers will have the experience to decide the basis on which they employ casual labour. For others the choice is to offer short term formal employment, with all the paperwork involved, or only to employ those who are self-employed workers in the building industry. The latter course is usual, and it is important to recognise this in order to meet any challenge from the PAYE authorities or involvement with the Employment Acts. This can only be a loose business at best, but a note from a duplicate book confirming the arrangements made will help avoid confusion. It might read

> 'Agreed. J. Smith, self-employed, general labouring during last week July only, £2.25 per hour, hours as required only. Not a contract of employment. XYZ'

Finally, a job programme is essential. This can be an elaborate bar chart showing critical paths or simply a list written down in a pocket book. But it must exist, be realistic and be used. The value of a job programme is not simply that you know when a particular job is to be done but is an aid to ensuring that all the tools, services and other requirements for the job are ready when needed. Every job programme is different, but a typical simple work sequence for a bungalow on strip foundations with a solid floor is set out in forty stages on page 62. The programme can be elaborated, or can be used as it is. Starting on page 64 is a sequence of photographs showing what this means on the ground.

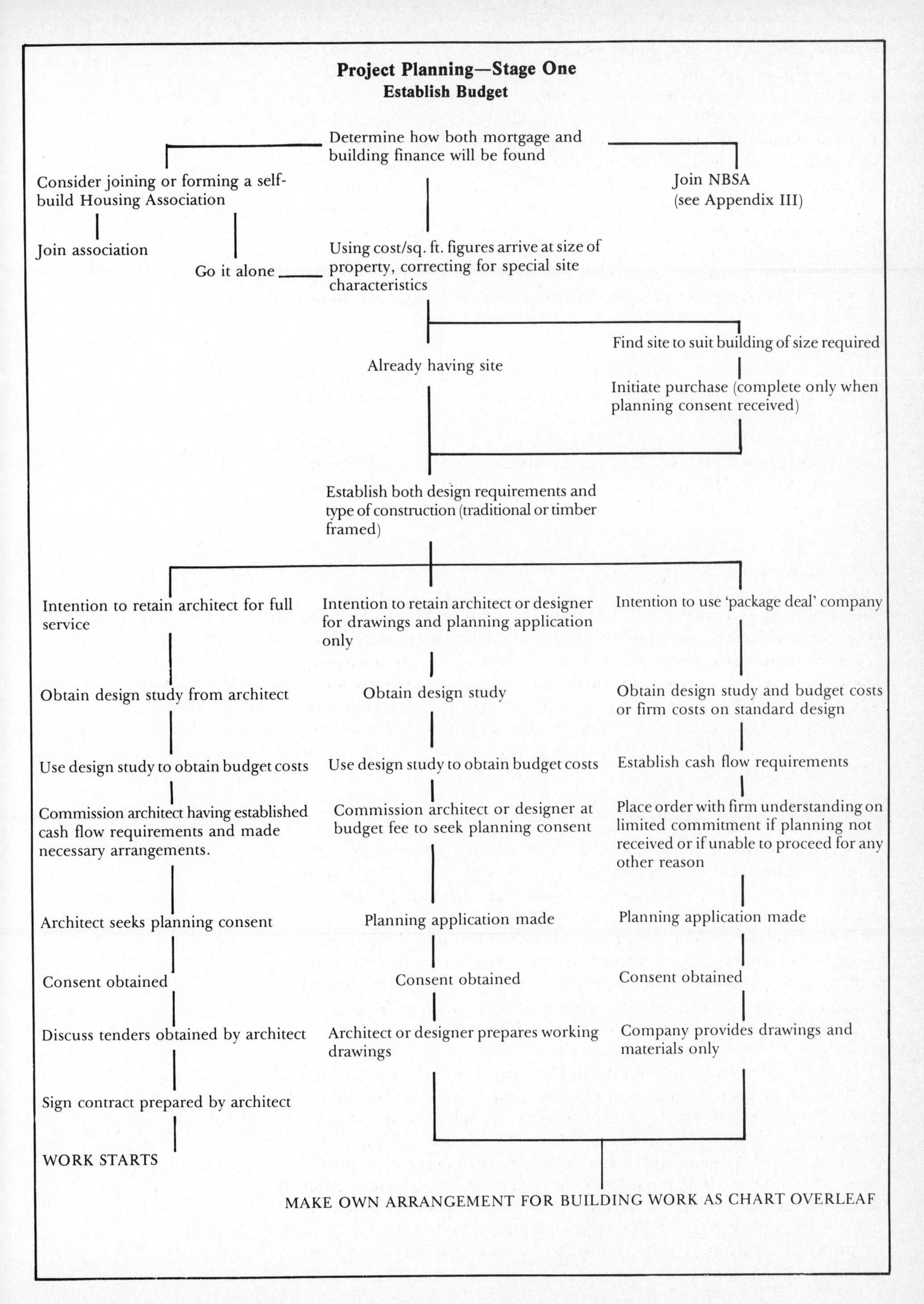

Project Planning—Stage One
Establish Budget
Determine how both mortgage and building finance will be found
Consider joining or forming a self-build Housing Association
Join NBSA (see Appendix III)
Join association
Go it alone
Using cost/sq. ft. figures arrive at size of property, correcting for special site characteristics
Find site to suit building of size required
Already having site
Initiate purchase (complete only when planning consent received)
Establish both design requirements and type of construction (traditional or timber framed)
Intention to retain architect for full service
Intention to retain architect or designer for drawings and planning application only
Intention to use 'package deal' company
Obtain design study from architect
Obtain design study
Obtain design study and budget costs or firm costs on standard design
Use design study to obtain budget costs
Use design study to obtain budget costs
Establish cash flow requirements
Commission architect having established cash flow requirements and made necessary arrangements.
Commission architect or designer at budget fee to seek planning consent
Place order with firm understanding on limited commitment if planning not received or if unable to proceed for any other reason
Architect seeks planning consent
Planning application made
Planning application made
Consent obtained
Consent obtained
Consent obtained
Discuss tenders obtained by architect
Architect or designer prepares working drawings
Company provides drawings and materials only
Sign contract prepared by architect
WORK STARTS
MAKE OWN ARRANGEMENT FOR BUILDING WORK AS CHART OVERLEAF

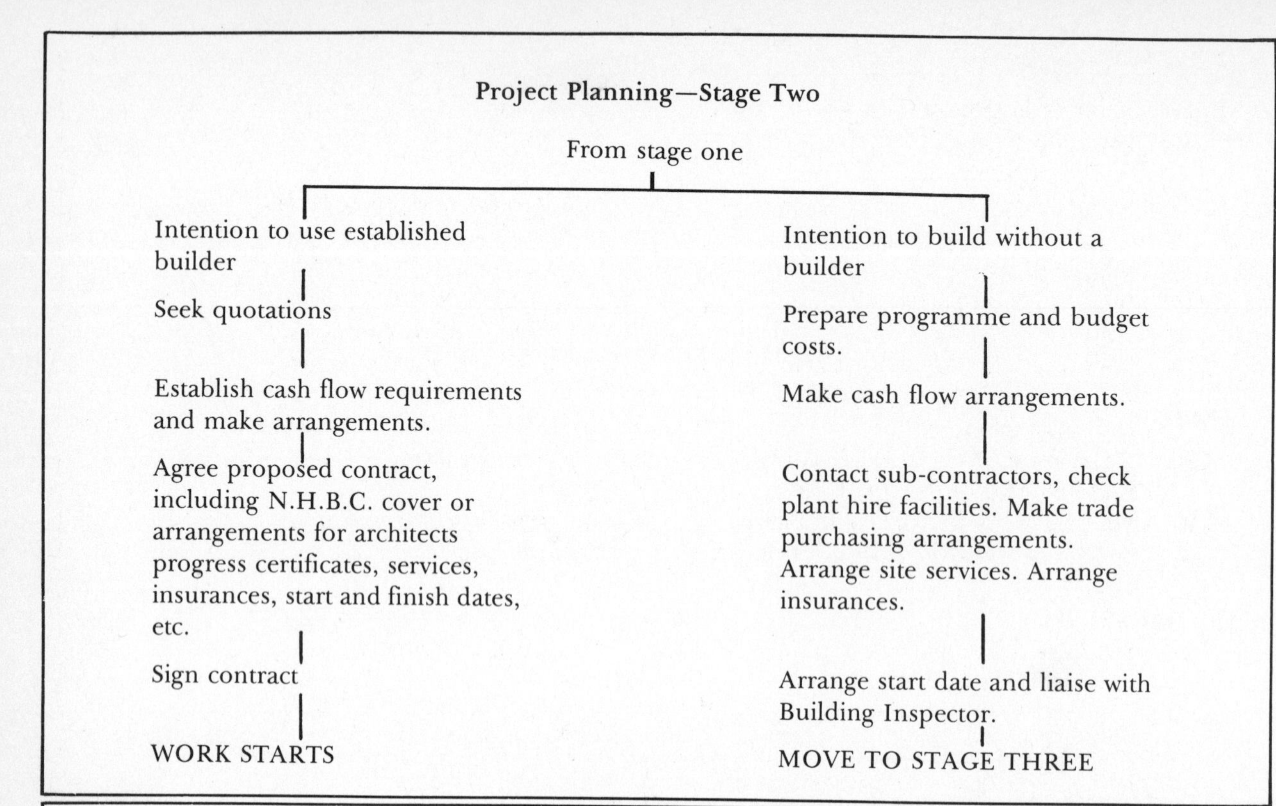

Project Planning—Stage Three

Typical Work Programme for A Bungalow On Strip Foundations, Solid Floors, Set Out In 40 Stages

	Action	Required
1.	Apply for all services.	
2.	Apply for building water supply.	
3.	Arrange insurances.	
4.	Provide access, hard standing, site storage, notice for deliveries.	*Hardcore, site hut, wire, notice.*
5.	Strip top soil and stack for future use.	*Digger hire*
6.	Excavate foundations. Excavate service trenches. Excavate drive and spread hardcore.	*Digger and hardcore.*
7.	Buildings Inspector to inspect footings.	
8.	Pour foundation concrete.	*Concrete. Possibly reinforcing materials*
9.	Building Inspector to inspect foundations.	
10.	Build foundation brickwork.	*Mixer, bricks, cement, sand, ties.*
11.	Fill foundations, blind, lay membrane	*Hardcore, roller, membrane.*
12.	Building Inspector to inspect foundation brickwork and membrane.	
13.	Cast slab, leaving ducts for services.	*Concrete, ducting for services.*
14.	Lay drains, fill soakaways, lay services if possible.	*Drainage materials, materials for inspection chambers.*

15. Building Inspector to inspect drains. Water Board and others may wish to inspect services.	
16. Build off dpc.	*Cement, sand, bricks, blocks, tie irons, insulation slab.*
17. Building Inspector to inspect dpc.	
18. Build up to square.	*Walling materials including joinery and lintels. Scaffold.*
19. Carpenter to scarf joint wallplate. Bed wallplate. Rear end trusses as template for gables.	*Wallplate and roof trusses.*
20. Build gables. Build in gable ladders. Build chimney up through roof.	*Chimney liners.*
21. Fix roof ready for tilers with barge and fascia fixed to suit type of tiles. Board valleys and make provision for flashings.	*Balance of roof materials.*
22. Sub-contract felting, laths, tiling. Plumber to be in attendance to attend to flashings and valley, or more usually tiles left unclipped for him to follow.	*Tiling contractor.*
23. Fix guttering required.	*Rainwater goods.*
24. Carpenters first fix — door linings, window boards, studding partitioning roof access. Hang external doors.	*Carpenters first fix materials.*
25. Glaze all windows, making provision for ventilation.	*Glazier or double glazing units.*
26. Plumbers first fix.	*Appropriate materials.*
27. Heating installers first fix.	*Appropriate materials.*
28. Electricians first fix, including telephone ducts.	*Appropriate materials.*
29. Plaster out.	*Ceiling board, plaster, angle beading.*
30. Lay floor screed.	*Insulation for piping under floor screed.*
31. Complete drains, external works, paths, drive. (Building drying out).	*Appropriate materials.*
32. Plumbers second fix and water on.	*Appropriate materials.*
33. Build fire-place.	*Appropriate materials.*
34. Electricians second fix, including TV aerials, and electricity on.	*Appropriate materials.*
35. Carpenters second fix, including kitchen and all fitted furniture.	*Appropriate materials, kitchen units and fitted furniture.*
36. Telephone installation.	
37. Decorations.	*All decorating materials.*
38. Lay tiled floors.	*Tiling contractor or tiles, adhesive, grout.*
39. Wall tiling.	*Tiling contractor or tiles, adhesive, grout.*
40. Clean through. Arrange for householders insurance to take over from site insurance.	

Tackling the Job

These photographs showing the progress of a self build project from start to finish were specially taken for this book with the help of all concerned.

Ken Andrews has built on one of the sites which the Chesterfield Borough Council make available to local residents who want to build for themselves. He is a self employed heating engineer and has always had a vague intention to build his own home, and as a local resident, knew of the Chesterfield scheme. When he saw that a new release of plots was being made, he contacted the local council, and was given every help to make a choice and to sort out exactly what he wanted to do. He arranged to buy a corner plot at the entrance to a cul-de-sac and promptly sold his own house and moved into temporary accommodation to release the money to get things under way. A building society mortgage was arranged but without progress payments. These were provided by the bank, who required architect's progress certificates in the same way as a building society.

Mr. and Mrs. Andrews went to Design and Materials Limited on the recommendation of others, and settled the details of the design very quickly. Planning consent was granted without delay, as only the design had to be considered. Building Regulation approval was given after lengthy consultations regarding a thin seam of coal just under the surface. A raft foundation was required to allow for this.

Work started on the raft a week before Christmas and the whole job took ten months. Progress is illustrated in the photographs.

Project Planning—Mr. and Mrs. Andrews visit the D & M office on the recommendation of a friend to discuss their requirements with one of the company architects.

Their ideas are incorporated in a design study sent to them a few days later. This design study is used to establish budget costs, and is the basis for the Planning Application made on their behalf.

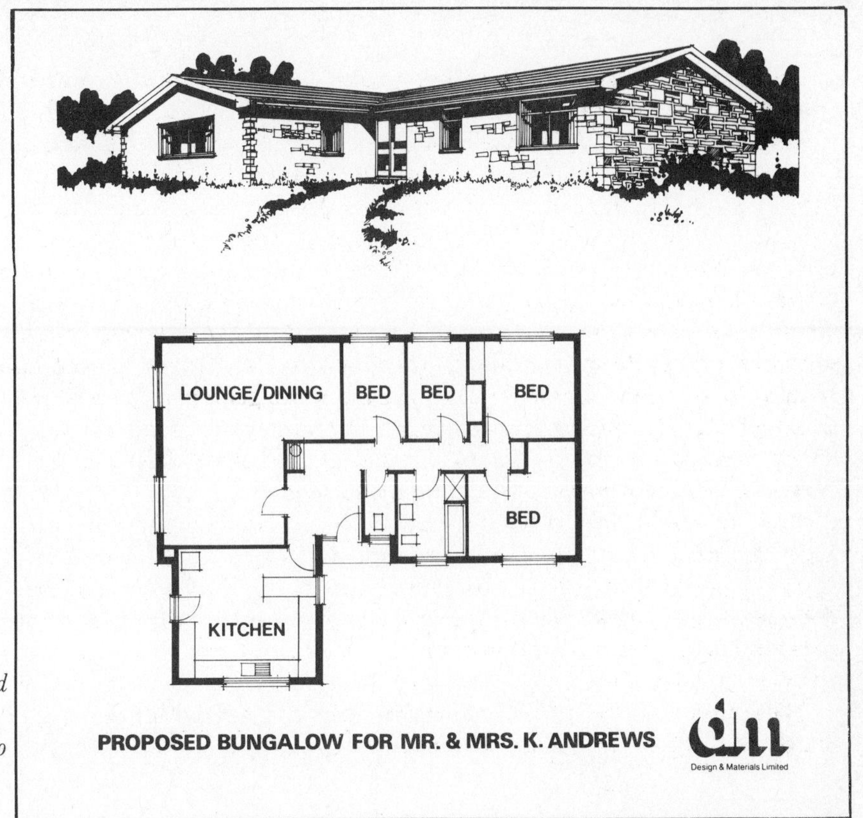

The site is in a subsidence area and the council require a raft foundation. The Andrews decided to construct this themselves, starting work on site in the winter. Here Ken and friends position the reinforcing in bad weather.

Ready for the concrete which is poured in even worse weather.

The finished raft. In the background are materials delivered by D & M Ltd.

Sub-contract bricklayers make a start on the walls which are to be built of reformed stone which is a perfect match with the local Derbyshire limestone. At this stage they are laying common bricks.

Ken Andrews did general labouring when required to push the job forward. He is a heating engineer and did all his own plumbing besides his central heating installation. He also did the plumbing for a friend building on another plot, and in exchange his friend laid the drains for both new homes.

Approaching wall plate level, with the hired scaffolding in position. While this was going on the services were being laid—drains, water, electricity, gas and telephone cables.

Fixing a truss as a gable end profile.

D & M Ltd., lath, felt and tile the roof as part of their contract.

The roof is tiled and all is weatherproof. The sacks on the roof are to protect the new ridge tile pointing from frost.

The first Christmas in the new home. A general view of the lounge, showing stone feature walling, but with the pelmets still to be fitted. The strip lighting to be fitted behind them can be seen.

The fitted kitchen exactly one year after work started on site.

The finished bungalow with the garden turfed.

Self-Build Housing Associations

About 7 per cent of owner-built houses are erected by self-build groups, sometimes with a great deal of support from Local Authorities.

A self-build group is formed by a dozen or so individuals who form an association to work together to build as many houses as there are members. When the houses are built each member moves into one of the houses, and they then wind up the association. By working together they can pool their skills, and by forming an association they can obtain loan funds and help in finding land. However, the group does not receive a grant or subsidy, and when the houses are built they have to be paid for by the members at cost, usually through a Building Society mortgage. The attraction of self-build is that this cost is very low. The attraction of group membership is that it provides the framework within which to build for oneself.

Self-build housing associations fall into two types, the managed and the self-managed. The difference between the two becomes more marked as one moves closer to the practical business of building. Managed groups are usually formed by an agency to build houses, already designed, on a site already purchased, using building finance already arranged. Members are chosen from the agency's waiting list or are sought by press advertising and can expect to be working on site within a month or two of joining. The agencies may be philanthropic, in which case the association is expected to provide its own site management, or commercial, in which case the agency expects to be retained as managers to the association on a fixed fee basis.

Self-managed groups have to find their own land, arrange their own legal formalities, and seek their own building finance. This is invariably a lengthy process as will be seen from some of the case histories. The delays and frustrations make demands on the enthusiasm of members and considerably extend the overall life of the association.

However a group is formed, the legal framework in which it operates is the same. It first registers under the Friendly Societies Act, 1965, with a constitution based on the model rules of the National Federation of Housing Societies, and affiliates to the Federation. To do this it has to have at least seven members, all of whom normally take up a £1 share. Sometimes a Local Authority or County Council will also take a share in the association. This gives associations a legal status, the ability to enter into legal obligations as an association, and limits the liability of members. It also gives an opportunity to seek loan finance, and assistance in finding land from the Housing Corporation, Local Authorities and others. Any attempt to build as a group without being a properly registered Housing Association would be most unwise: apart from any other consideration, a legal position is likely to develop where any one member could be responsible for the debts of all members.

When the association has been formed it elects its officers and a committee, and adopts the standard working regulations recommended by the Federation. These regulations set out in straightforward terms exactly how the association is to operate and what the obligations of the members are. They deal with matters such as the hours which members are expected to work, the ordering of materials, keeping accounts and the authority of the committee.

When the working regulations are agreed in detail each member signs a separate contract with the association to signify his agreement to them, and to

Members of the Ebley Self-build Association in South Wales celebrate the end of 14 months of hard work. Each member's initial outlay was £100, loan finance was arranged with the Nationwide Building Society by Manager Chris Biggs, and the final cost/value differential was 43%.

signify that the association will transfer a house to him when all the properties are finished. Before signing these contracts the members will usually have allocated the houses among themselves, often by ballot. Naturally not all the houses are finished at once, and at this stage it is also usual to agree the order in which they will be completed. All the new homes remain the property of the association until the last one is finished, and when members move in they do so on a formal 'licence to occupy' issued by the association. This does not constitute a legal tenancy, so that the association retains control over every house until they can all be transfered to members on the same date. This is important in many ways, not least to ensure that members who have moved into a home have an incentive to keep working on the other houses! Finally, the association is wound up, and remaining assets divided between members, together with a refund of the £1 membership share.

Model rules will be found at the end of the book. It must be emphasised that these should not be copied for us by an association and that the Federation should always be asked to provide drafts incorporating the changes made necessary by new legislation or other factors.

The financial framework within which the housing association operates is equally specific. It must appoint a treasurer to handle its accounts, receive monies and make payments, and to keep proper books of accounts. A professional accountant has to be appointed as auditor. Insurances and a fidelity guarantee have to be arranged. VAT registration has to be applied for, and an application made for the scheme to be approved for option mortgage subsidies. All this is arranged for a managed association, while the self-managed association must look to its committee to handle this work, using the publications of the Federation to assist it.

Both types of association obtain their finance in the same way as follows:

SHARE CAPITAL is the nominal £1 shares taken up by members.

LOAN CAPITAL is the contribution which members make to get the association going and to demonstrate their intention to support it. Loans are interest free, and are repaid or set against the cost of the house at the end of the scheme. These loans are made in cash as a condition of membership, or are paid as a weekly subscription. Associations sometimes require comparatively high loans from 'unskilled' members and waive loan obligations from members who are skilled tradesmen.

BUILDING FINANCE This is the money to get the job done, probably nearly half a million pounds which has to be borrowed until it can be paid back by the members individual mortgages when they move into their new homes. Self-managed groups often get their loans from commercial banks with the risk underwritten by merchant bankers Morgan Grenville using a fifteen million pound Housing Corporation revolving fund. In 1983 this fund made loans for 231 houses, and received re-payments from mortgages on 312 houses which had been finished. Managed groups also use this fund, or they may negotiate outside commercial finance, often from a consortium of two or three banks or building societies. A local authority loan is another alternative, particularly if the local authority has found the land for the group. Remember that all of this building finance carries interest charges which are a major element in project costs.

Having discussed the formal arrangements for housing associations it is time to look at the members. Here one would expect flexibility, with scope for individuals to get together simply as people with a common aim. Unfortunately this is rarely the case. Housing associations are afforded a remarkable degree of support by various bodies, but in exchange, they must be established in a way which experience has shown to afford them the best chance of success. The ideal membership is between 12 and 20 in any one group. The Federation is specific in this, as is the Housing Corporation and others who provide building finance. They look for a group which has 50% of its members with building skills, such as bricklayers, carpenters, plasterers etc., to give a professional feel to the operation. They expect the majority of members to be married and to possess a steady outlook that gives confidence that they will stick a year of unrelenting toil. They look for a balance between men over thirty and under thirty. They look for a membership established in their employment and therefore good mortgage risks.

Managed associations carry this to extremes, making up lists of possible members from applicants as if they were selecting an England football team. Few employers would admit to being as selective, or looking so carefully at all the circumstances of a prospective employee, but this rigorous selection has one justification — it works. Building one's own home is working outside the system, and needs special people. Unless everything is right, the odds against success will lengthen. The Federation, and the experienced professional managers, do know what makes a success. Their advice should not be lightly discarded.

Individual members of groups contract to provide loan finance as agreed, arrange to buy their house from the association at the end of the scheme, and to work for the association as set out in the working regulations. Typically, members agree to work a 20 hour week in the winter and a 28 hour week in the summer, plus one week of their holiday. Seventy five per cent of working hours must be worked at weekends. Often experienced bricklayers, joiners and plasterers receive an allowance of two free hours a week, recognising their higher productivity. Special arrangements cover sickness, otherwise absence or lateness results in an automatic fine, often £5 per hour. This is not paid in cash, but is built into final cost of the member's house. These fines are rigorously imposed by most associations, who feel that only in this way can

Group Self-build at the very top of the market. The Rectory Self-build Housing Association built 14 houses like this at Two Mile Ash near Milton Keynes.

personal relationships survive resentment against the offender.

All members are expected to specialise in one aspect of the building work as decided by the committee, with all participating in tasks such as concreting, path laying and unloading materials. Experienced building tradesmen have their own role, the inexperienced quickly acquire skills, and specialist work is put out to contract when necessary. This is particularly important where unprofessional work carries a risk—gas fitting, electrical work, scaffolding etc.

Occasionally associations lose members due to ill health or other reasons, or expel members who default in their obligations. The rules make provision for this, and associations invariably arrange for a new member to fill the vacancy. The member who leaves takes the money which he has put into the association but usually forfeits the value of the work which he has done. The Federation advises all associations to arrange life insurance for members so that the association can be paid for their lost labour should they die, and their dependants can then take over the house.

At this point, the roles of the sexes are of interest. There is a clear-cut pattern which is impossible to explain in simple terms. The man building for himself on his own invariably works in close and effective partnership with his wife. Their joint involvement in the new home is complete, and invariably they speak of what 'we' are doing. If the wife plays no part on the site she will be organising materials, chasing sub-contractors, deducting monthly payment discount from the bills. She will discover unsuspected tenacity in dealing with the water authority, gas and electricity boards, and British Telecom. Frequently she provides far more than half the enthusiasm and drive required to carry the job through. Often she will work on the site with her husband and sometimes the only visible difference between them is that the wife wears gloves. Such wives terrify architects.

In group projects, surprisingly, wives play virtually no part. Sometimes one hears of a woman secretary ot treasurer but this is rarely reported as a success. They seldom visit sites except as visitors or to make tea, and some associations restrict family visits to specific times. The exception is where associations leave interior decoration to members and this is done by the wives while the husbands are working on other houses. I make no attempt to explain this phenomena and simply report the relative roles of the sexes as they are seen on the different types of site. In the case studies later, I have quoted those who point out that membership of a self-build association is hard on a wife who may expect to see little of her husband for a year or more. The inference is that it is hard on the marriage. Individual self-builders never have this complaint. All the individual self-build couples quoted here were interviewed together, while the group self-builders did not call their wives to join in the discussion.

Self-building as a member of an association has only two disadvantages. Firstly the design of the group house is agreed by the group as a whole, and once settled there is no opportunity for individual variations. It would seem a simple matter for groups to arrange for variations to different houses, but in practice this presents problems, and both the Federation and the management organisations advise strongly against it. Colour of bathroom fixtures and the selection of kitchen units are normally a matter of individual choice, but little else. The standardisation and personal choice is much the same as that on a developer's housing estate, and this may not always suit the independent spirits who want to build for themselves. The other potential disadvantage of building in a group is the constraint of the group discipline and the need to work to a set programme. For some the team spirit which this fosters is one of the attractive features of the whole business, but for others it presents problems, particularly for building tradesmen who sometimes find it difficult to accept that attitudes on a self-build site are different from those to which they are used in their everyday work. However, these constraints of philosophy and outlook are really side issues.

Members of the Malling Self-Build Housing Association celebrate finishing their new homes.

Management Consultants

There are at least a dozen management consultancies offering services to self-build associations. To some the existence of professional management in the movement seems sacrilegious, and the idea of groups of dedicated do-it-yourselfers paying others to direct their labours certainly does take getting used to. Perhaps this reflects a particularly British approach to management, for it is obviously appropriate for amateur builders to employ professional solicitors, architects, surveyors or accountants, and the management skills required to put together a viable self-build group, and to ensure its success, have to be of a very high order. Be this as it may, professional management on a commercial basis is an important feature of the group self-build scene, and accepted by all who have influence within the movement. This acceptance received a setback in 1980–81 when a new Management Company advertised widely, formed a number of associations, and subsequently had accusations levelled against it by many of its member associations. The company concerned is not now active, but established management companies, worried by this, formed the National Society of Self-Build Consultants and drew up a Code of Conduct. The Society is not now active, but its Code of Conduct, with only minor modifications, is now recognised by the Housing Corporation and the Federation of Housing Associations, who will both advise whether a consultant is known to them as having accepted the code.

Of the Management Consultants, the group of seven companies linked with Wadsworth & Partners of Bradford, is pre-eminent. It has been responsible for 210 completed schemes totalling over three thousand houses in fifteen years, with 49 associations currently building 760 new homes. The Federation handbook on self-build contains whole sections reproduced from the Wadsworth publications, and acknowledges this in its opening pages, and the companies liaise closely with the Housing Corporation and many other quasi-government bodies. The reason for this special position is simply that the self-build movement needs the Wadsworth Group even more than the group needs the fees which it charges to its client associations.

In every analysis of the performance of a self-build group, professionally managed or not, it is seen that successes or failures are determined before a brick is ever laid. The right site, the right design, the right contracts for roads and drains, the right finances, the right liaison with a score of authorities—all these are the prerequisites of success, and without them the problems will multiply. Guaranteed success requires an expertise that can only come with experience, and it is this experience which managed associations are happy to pay for.

The Wadsworth companies offer management services in all parts of the country, with their clients coming to them in various ways. They are sometimes approached by an embryo group seeking a management service, or by groups experiencing difficulties, but their usual pattern of operation is to set up the infra-structure of an association and then advertise for members. This is a pattern of operations common to all the dozen or so self-build consultants, and it is in establishing a firm basis on which an association can be formed that the company really earns its fees. It has to locate a site, obtain an option of some sort on it in the name of an association that does not exist—no mean feat this—obtain an architect's draft development proposals, prepare a costed scheme based on these proposals,

Self-build consultancies are big business, and may use well known T.V. personalities to help launch new schemes. Here Bernie Winters and Snorbitz open an exhibition by Wadsworth and Dring in the W. Midlands. However, in spite of the hype the consultants always emphasise the blood sweat and tears as well as the rewards.

obtain the promise of finance for the proposals from a bank, building society or the Housing Corporation, obtain planning consent, and finally obtain members. All this preliminary work is rarely possible without the co-operation of a Local Authority who have to be persuaded that there is a real need for a self-build group scheme in their area.

When a blueprint for a viable scheme emerges, most consultants arrange local advertising for a public meeting, at which they present the opportunity to join the association, and invite applications for membership. Average attendance at such meetings is about 200, although numbers vary widely in different areas for no discernible reason. At the meeting there is a display of photographs of other schemes, literature is available, and the hard facts of the proposals for the new association are set out in an information sheet that shows the design and quotes the anticipated finished cost. It is all specific, very hard on facts and very low key on salesmanship. Those interested are asked to fill in an application form and pay a nominal deposit. The fees that the association will pay to the management company are explained, and these typically will be 7½% of the value of the properties, although there are wide variations. In presenting schemes to prospective members the management fees are always included in the quoted target costs. For these fees, associations receive a service that is carefully defined in a formal contract with the managers and which is a requirement of the working regulations. This contract defines the managers' responsibilities which are usually as set out on page 77.

As the land will have been obtained and the house design settled before the association is formed, it is necessary for the solicitors and architects employed to be nominees of the managers, although independent of them. One advantage of this is that the solicitors will be used to the legal problems of housing associations, and with working with the legal departments of the Housing Corporation and the Building Societies. As with much else in

Typical Professional Managers' Duties

To advise on the overall organisation of the association, and to attend to all its legal affairs.

To deal with the acquisition of land, and all the easements, services and other arrangements that have to be made.

To deal with the architect regarding design work plans and planning consents.

To provide a price schedule of material for each house design and to arrive at the actual cost ruling at that date.

To negotiate finance with the Local Authority or Housing Corporation to finance the scheme.

To arrange repayments to be made to the authority at the end of the scheme, and to help members to obtain private mortgages.

To keep the correct books of account for the association, and to prepare monthly statements of account for it.

To issue loan stock as required and keep personal ledgers for each member.

Deal with VAT affairs.

Prepare the final accounts, dealing with matters such as stamp duties, legal fees, mortgage transfers fees, interest charges, etc.

Obtain tenders for the roads and services, and devise the association on the contract and supervise work.

Obtain prices and negotiate arrangements on behalf of the association for sub-contract labour, plant and machinery etc.

Obtain surveyor's certificates relating to the standard of the buildings as will be required by members for their mortgages.

Provide a contracts manager to maintain a close and effective liaison with members regarding the work in progress, to see that properties are properly built, and to offer advice and assistance where needed.

Keep the association advised regarding current and projected movements in prices and building costs, and to assist them in dealing with this situation.

Attend most of the monthly meetings of the association's committee. Deal with final habitation certificates and licences for members moving into a house before the end of the scheme.

Generally act as advisers to the association in all technical matters.

(Courtesy of Wadsworth & Partners)

managed groups, this lack of choice has to be set against the advantages of using the experience of others.

Some consultants have recently introduced a new type of association where the members co-operate to build the shells of their houses but each members makes his own arrangements to complete his home, using the group's facilities as he requires. It seems likely that this approach will be further developed, and may lead housing associations further away from the low-cost housing than those who founded the movement envisaged.

Finally, anyone in doubt about the standing of a Building Consultant or management company offering help to associations should seek the advice of the Housing Corporation or the Federation of Housing Associations; the addresses are in Appendix III.

Self-Build and the Local Authorities

The attitudes of local authorities to D.I.Y. builders are diverse and reflect special circumstances rather than any overall policy. The official position is that Councils have been urged by successive Governments to consider helping self-build Housing Associations, and a Department of the Environment Circular issued in 1975 lays a duty on Councils to examine whether they can find land for local associations, and consider whether they should publicise that they have land available for this purpose. To say the least their enthusiasm for doing this varies.

Some councils have a long tradition of supporting Self Build groups, some favour self managed groups and others prefer to have professional managers set up schemes. The role of a council can vary from simply selling land to an association at market value, as an ordinary disposal of land made in the usual way, to making special arrangements to find an association land that would not be available to others, and lending the association advice and support at all stages. This sometimes extends to the Chairman of the council's housing committee taking the chair at a public meeting to form an association, and this sort of support is invaluable.

Other councils appear to be wary of groups, but have an active policy of releasing serviced plots to individuals, many of whom they know will build for themselves. Some local authorities will make a point of advertising that such plots are suitable for self builders, and this simple recognition of the self build approach is usually enough to ensure that 75% of those who buy the plots will build on their own, while the rest will employ builders. The way in which such plots are allocated varies, with many councils giving local ratepayers the first chance to buy, and only selling to outsiders when the local demand has been met. Sometimes these plots are offered by tender, sometimes allocated at fixed prices from a waiting list, and occasionally allocated on a first come first served basis on a particular day, which can lead to applicants queuing for a week. Oddly enough the queue system is popular with purchasers, who turn them into week long pavement parties, and a good way to get to know each other. In Milton Keynes, where they were an established feature until recently, both architects and builders merchants could be seen discreetly soliciting business among the queuers!

Whether or not a council sells land in this way, and with what enthusiasm, has little to do with political attitudes. To start with the council has to have the land to sell, and if they have not it is unusual to hear of land being bought by an authority simply to make it available for a self build scheme. Generally speaking, councils that sell plots are in areas where development is being actively encouraged, and this is not in the Home Counties or in the South East. The only generally available information about what is on offer is to be found in a quarterly leaflet published by Ryton Books of Worksop which they distribute free of charge to those who buy their books, including this book, by mail order.

In succeeding editions of this book many different local authority schemes have been examined, and in this edition we look at the Sutton in Ashfield District Council in Nottinghamshire, where in 1985 the Estates Department sold a development of 47 serviced individual building plots in a village called Underwood, together with a site for a self build housing association. To round off a carefully planned scheme a final plot for a corner shop is available for a self-build-shopkeeper. The Housing Association concerned is featured in the case histories on page 128, and there are photographs of the site as at November 1985 and February 1986 below.

The Underwood project was presented in a forty page booklet which was available from the council early in the year. The booklet contained application forms, and every enquirer got two copies, so that one could be returned with the application form completed. In this way no one could claim that they did not know the rules! Everything was set out very clearly, in six chapters, introduced as follows:

1. Details of plots, purchase price, conditions of sale
2. Design brief to be followed
3. Site development arrangements, services, arrangements to avoid inconvenience to others
4. Finance. A note on common forms of finance with details of special terms offered to self builders by the Trustee Savings Bank.
5. Purchase procedure
6. Application form.

The plots were allocated on a first come first served basis on April 10th, with local residents given preference for seven days. A long queue started forming on April 7th, and all plots were taken up by Sutton in Ashfield residents. Within six months the first homes were occupied, and it soon became obvious

Melvyn Butler, of the Sutton-in-Ashfield Estates Department talks to a self builder at Underwood in November, and the same view only three months later as the first homes are approaching completion.

that the whole development was going to be very attractive. As a result, house prices elsewhere in the village began to move up very quickly, to the delight of all concerned.

The Ashfield District Council has not imposed any conditions in the title to restrict the resale of the houses. In the past it was common for an authority to include a pre-emption clause in contracts for the sale of building plots to individuals, but this practise has now ceased. A proportion of self-built homes are sold after they are built, although the family concerned will invariably have moved in first to avoid capital gains tax. This leads to speculation that self builders are often building with the intention of selling their new house and taking a profit. This is certainly over emphasised, and it must be remembered that those who build for themselves are among the most lively and mobile section of the population who expect to change jobs and homes more often than the average.

Recently the lead in all this has passed from Councils to the Development Corporations responsible for the new towns, and it is interesting that in terms of houses built they have had more success with individual self-builders than with self-build associations, although they are active in both fields.

Milton Keynes in Buckinghamshire plans to build 20,000 houses in the 1980's, grouped in 'villages' each with its own services and its own distinctive character. Like all modern new towns, it is trying to avoid a 'council estate' image. Most of the new homes are built by national developers for sale on the open market. There are co-ownership schemes, modern mortgage arrangements of all sorts, and a special effort has been made to accommodate the needs of self-builders. At the time of writing the Development Corporation has marketed 800 plots, ranging from less than a tenth to a third of an acre, all fully serviced with tree vouchers to help with landscaping! Waiting lists are maintained, and most plots are sold by tender although some are advertised at set prices. Releases of these fixed price plots have led to week long queues outside the Housing Offices in the past. Six sites have been sold to self-build groups and more are in the pipeline.

Building at Milton Keynes is certainly different: here there is a bureaucracy determined to make things happen. The ordinary planning procedures do not apply, and under special provision of the 1965 New Towns Acts, plans for new houses are considered by the Development Corporation's architects across the counter, with no formality and no delay. This does not mean that design requirements are not precise and strictly enforced, but everything is settled quickly and easily.

The Corporation has a 'Build now—Pay later' scheme whereby a £100 deposit and only 10% of the plot value will secure a site. The balance of the cost, plus interest from the date that building starts, can be paid when the house is completed. This is invaluable to self-builders who can put their own money into bricks and mortar, and helps the problem of paying for the materials required to take the building to the stage when the first payment is received from the Building Society. The Corporation maintains a Home-

SELFBUILD OPPORTUNITIES

As described in the text, the only available information about opportunities to become a self builder is a list of plots for sale and self-build groups seeking members, published by Ryton Books, of Ryton St, Worksop every quarter. It is sent free of charge to those buying the companion volumes to Building Your Own Home using the order form on page 179, or send 50p to cover printing and handling. The Spring 1987 leaflet had brief details of 84 new self-build associations and 32 councils with land for sale.

finders' Centre to assist prospective purchasers and advertises its services widely.

Both plot cost and house values tend to reflect Home Counties prices. The value of homes built on these plots go up to over £100,000 and there is a lively market for them. Tim Skelton, who handles the land sales at the Corporation's Housing Unit, estimates that 10% of the plot purchasers use registered builders, and 90% use sub-contractors together with their own labour. The latter make the usual saving of at least £5 per sq. ft., and claim savings of at least £10,000 by building for themselves. They probably have less worries than self-builders elsewhere, because they build in a community where everyone else is building. Advice, encouragement, optimism, criticism and counsels of despair are all available from neighbours.

Both Councils and authorities like Development Corporations are now running more schemes of this sort than ever before, but it is uncertain whether this is due to a desire to help people build their own homes, or whether Council Treasurers are insisting that their Estate Departments turn surplus land into cash to meet a financial crisis! However if a Council has experience of a successful scheme a follow-up scheme is more likely.

Self-build housing at Chesterfield in Derbyshire.

Inner City Self-Build

Group self-build in its modern form started in Brighton in 1948 with special schemes for ex-servicemen to build their own homes on land provided by the Borough Council. They were a success,* and as such schemes proliferated in an era of controls and nationalisation they attracted grudging recognition, limited support, and a concern that they should be fitted into the housing bureaucracy. By the 1970's a pattern had emerged where they were usually financed by the Housing Corporation, used the National Building Agency as consultants, and were expected to register with the Federation of Housing Associations and to adopt its model rules. The land upon which they built was usually made available by a local authority, which often imposed pre-emption clauses in titles to prevent self-builders selling their homes to make a quick profit. There were other approaches too; in the 60's and 70's the West Midlands Council of Housing Associations financed 2,000 self-built homes in Birmingham, often using funds from Quaker charities. In time all of this became formalised, and so did the Housing Corporation's view of self-build associations: they were specifically for those in housing need, defined as being those on a council housing list. Associations with any significant proportion of members who were not on housing lists were not eligible for housing finance, and found it very difficult to obtain money from other sources. Self-build associations were very definitely considered to be a way in which the government helped those in housing need to help themselves.

In the late 1970's all this started to change. Financial stringency had closed the NBA and severely limited the finance available through the Housing Corporation. The self-build management consultants had arrived and had found sources of private finance for the associations that they managed. Within a decade the typical self-build home was a four bedroom, detached unit instead of a three bedroom semi, and 'housing need' was rarely mentioned where self-build was at its strongest, which was, and still is in the outer suburbs and small towns. In the late 1970's self-build had abandoned the inner cities where housing needs were still very real, and had become something for those who were using it as a way of jumping several rungs up the housing ladder at one bound. It had little relevance to those who wanted to get a foothold on the bottom rung unless they were exceptionally determined or able.

All things move in cycles, and today there is a new interest in self-build for those in housing need, and this is likely to become very important again, although the number of homes involved at present is still small. In its most interesting form it is part of the Community Architecture movement.

Community Architecture is a phrase that means different things to different people, but in general it contends that many of the ills in our society come from ordinary people having been ignored when their "built environment" was planned and constructed. Today's inner city housing is seen to demonstrate this very clearly, and Community Architecture is particularly concerned with inner city rehabilitation areas.

A typical approach in this is to stimulate a local community to press for an urban rehabiliation scheme, and when this is approved to ensure that as many of the existing residents as possible participate fully in every stage of the planning, leading to them becoming involved in the actual building work in

*In all 1194 homes were built at Brighton by 58 associations over 20 years.

some sort of self-build format. Those who promote this approach to try to involve the whole of the existing community, and so far as building work is concerned their differing capacities for work restricts each person to working on his or her own home. The Community Architecture approach is seen as leading to a general improvement in the lives of all concerned, with new skills leading to new job opportunities, and wider social awareness. This is very

Dr Rod Hackney and Prince Charles at a rehab scheme at Stirling which included a number of self-built homes. Royal patronage of community architecture is one of a number of factors which is giving self-build a new high profile in the media.

different from suburban self-build with its ever-present concern with cost/value ratios.

There have been some particularly successful schemes run on these lines under the direction of Dr. Rod Hackney, the charismatic President of the R.I.B.A. He has arranged royal patronage for some of them, as seen in the photograph above, and his own projects have influenced many more.

Another approach to group self-build housing along these lines has evolved since the 1970's at Lewisham, where it is described as being the 'Walter Segal approach' after the innovative architect whose timber frame designs have been used by the groups concerned. In some ways this is confusing because the structure, management and financing of these groups is novel enough to merit close attention apart from the system of building, as the schemes are devised so that anyone can take part irrespective of their income, age or lack of building skills. However, as those who are concerned

A Segal house in Lewisham.

A Segal type house under construction, showing the simple post and beam construction which permits a very flexible arrangement of the internal walls.

with the schemes believe that their success is largely dependent on the merits of the Segal houses, it is appropriate to look at what is involved.

Walter Segal, who died in 1985, was an independent if not controversial figure who developed a modular timber frame system which can be used to provide one or two storey homes at very low unit costs. The essential features are a light post and beam frame, as illustrated in the photographs, capable of being made on the site, providing an uncluttered envelope which can be fitted out to suit the needs of the occupants. Lightness, both in the weight of the structure and in the size and arrangement of the windows, is very important, and the resulting appearance of most Segal homes is at variance with todays post-modern styles. Be this as it may, Segals designs have their own niche in the contemporary housing scene, and like those of other trend-setters they have an influence out of proportion to their numbers.

They are also very well suited to self-build projects, combining an opportunity to actually make the simple frames on site from sawn timber, and then to divide the interior into rooms to suit the life style of the family concerned. Each family can build their own individual house, with all the households cooperating on the common items such as drainage. There have been two such schemes sponsored by the London Borough of Lewisham, within an organisational framework that is as novel as the house designs. The present scheme of 13 units was started in 1985, should be completed in mid 1987, and embodies the experience gained on other projects. The land is owned by the Local Authority, who finance the whole of the building cost except for the labour, which is provided by the self-builders. The finished homes are then occupied on a shared ownership basis, with the council granting each self-builder a 99 year leasehold on the building, the cost of which is discounted by the value of his or her labour. The self-builders have an option to buy 10% units in the councils share of the property until he or she owns all of it. The usual arrangement is that the self-builders efforts entitle them to 50% of the home that they have built, the balance being rented from the council, and most families go on to buy the balance as and when they can.

None of the leases granted so far have been resold, and the cost/value equation which is usually studied so carefully is considered much less important than the real value of getting the individuals concerned into good detached houses which are designed to suit their individual needs and financial circumstances. Participation in the schemes has been limited to those on the Borough housing lists, and the social benefit of getting the members housed, of giving them a chance to participate in both the design and the building process, and to learn new skills and to gain in self-confidence is considered to justify the considerable measure of support provided by the Borough authorities.

Meanwhile in the London Docklands no fewer than five more conventional self-build schemes are under way in early 1987, and many more are in the planning stage. In Liverpool the various organisations linked with the local Cooperative Housing Movement have interesting schemes afoot involving different levels of community participation, while in Telford the Lightmoor project is putting the same philosophy into practise in a rural context — not an inner city scheme, but part of the same movement toward total involvement of a community with its 'built environment'.

All this still only accounts for a tiny fraction of the total of self-built homes, but it is important to all selfbuilders, as the media exposure which these schemes attract helps to make self-build fashionable with both councils who release building land, and with the financial institutions who provide development finance. In terms of rate of growth this is certainly where the action is, and it is hard to predict where it will lead.

The Great Eastern Self-build Association is building on the Isle of Dogs near to the slipway where Brunel built his famous ship. This large group of forty local people are building traditional houses following the normal self-build pattern, but with a pride in it being a community venture that is seen in the site signboard.

Energy Saving

No-one building a new home can ignore consideration of energy costs, nor are they short of advice on the subject. On every side we are urged to adopt new ways of reducing heating bills, and when building a new house there seem to be many opportunities to do this. Unfortunately the choices are bewildering, and it is hoped that this chapter will help to explain some of them. The first advice is to be suspicious of all advice. Low energy technology is trendy and important. It generates a mass of seemingly authoritarian literature which is often slanted, and may mislead someone who is building a new home and who wonders what arrangements should be made to ensure that his heating costs will be as low as practicable. It is said that more of the world's energy resources have been depleted by printing articles about solar energy than all the energy yet collected by all the world's solar energy collectors. The same is probably true of related subjects. 99% of this literature is put out by either those selling something, or by wild enthusiasts, neither group famous for careful evaluation and dispassionate judgement. Most newspaper and magazine articles are based on this literature, and are often written with one eye to a 'good story'. The same is true of T.V. features. Reflect that a T.V. feature to demonstrate that something is *not* the energy source of the next century would have few viewers, and beware taking 'Tomorrow's World' too seriously.

Moving on from this sterile scepticism, what can you do? Remember that new techniques and equipment cost money, and that this will be *your* money. If you are going to spend your money to save energy it will be for one of three reasons, and it is very important to differentiate between them.

- Either your interest in the matter transcends sordid financial considerations. Cost effectiveness is not important provided that the arrangement works.
- Or you wish to invest in arrangements that are not cost effective now, but which you anticipate will become so in the future because you expect fuel costs to increase ahead of inflation
- Or you only wish to incur expenditure on arrangements that are cost effective now, and which will give you an immediate saving of at least ten per cent of their capital cost per annum.

Decide where you stand, as different approaches are appropriate to the different viewpoints.

All new energy saving ideas are about arrangements which either keep heat from leaking out of a building, or offer a new and particularly efficient way of using energy sources, or involve design features to make use of solar energy in one way or another. No brief mention here can do justice to all of them, and the best way to study the subject is at a Building Centre, or better still at a specialist exhibition such as Energy World* at Milton Keynes. Keep in mind

* The Energy World exhibition at Milton Keynes involved forty different homes all built on the same site by many different developers and organisations to demonstrate every conceivable approach to energy saving in domestic construction. At the end of the exhibition the houses were sold and they are now mostly occupied, but a very informative handbook to it all is available at the Milton Keynes Information Centre and the homes can be viewed from the outside. It is important to see them if you want to make your own judgement of the appropriate balance between energy saving design and conventional appearance.

that many products are correctly promoted as offering cost effective fuel savings when used to improve a poorly insulated building erected thirty years ago, but are not applicable at all to a new building with built-in insulation to today's mandatory standards. It is important to disregard all advice on techniques applicable to existing buildings.

Starting with fundamental design considerations, houses that are designed to make the best use of the heat of the sun usually have a large area of glazing on the south side of the building through which the sun can heat a massive masonry internal wall called a trombe wall. This acts as a huge storage heater, sometimes using special air circulation arrangements. The limitation of this is that it assumes that it is convenient for the house to face in one particular direction, and that a conservatory or similar glazed area is aesthetically desirable and suits the lifestyle of the occupants. There are dozens of such houses of different types at Energy World, and you can judge for yourself how the planners are likely to react to one in the area where you want to live, and what sort of a resale potential it would have.

Insulation is a different matter, concerned only with stopping heat from escaping from houses of any style or design. The level of insulation which the Building Regulations require in a new house has been steadily improved over the years, and for 90% of situations the requirements are now that

- Floors need not be insulated, although it is likely that this will be required when the regulations are next revised
- Walls need to provide more insulation than is given by the traditional cavity wall with an inner skin of ordinary 4″ insulating blocks. This is most commonly arranged by putting insulation in the cavity as described later, and this need not involve losing the benefit of the cavity. Alternatively there are special blocks which enable a cavity wall to meet the standard without additional insulation
- Limits on the area of glazed windows in any one wall, with 12% only of the total wall area permitted to be single glazed, 24% permitted if it is double glazed, and 36% if triple glazed
- 4″ of roof insulation

Additional insulation can be specified by anyone who thinks it worthwhile to do so. The issue is whether or not it is worth the extra cost. The can only be evaluated for specific homes, heated with specific fuels in specific appliances, with the occupants wanting specific temperatures for a certain number of hours a day. In virtually all cases the extra cost of providing extra insulation is not immediately cost effective.

Insulation as an investment is a different matter, and it is very popular to build a new home with more insulation than the required standard. If you want to do this you can consider the following, which are roughly in order of cost

- Sophisticated draught seals to doors and windows
- Foil backed ceiling boards
- Additional roof insulation
- Triple glazing, or insulating glass which passes heat in one direction only
- Insulated floors
- Higher levels of wall insulation

All the above will have paid for themselves in twenty years, and probably long before. Whether this is the right investment to make is a question for the individual. Quite simply, a spare £1000 has many calls on it besides paying

The Llewellyn Homes house at Energy World requires only one third of the energy used in a home built to the Building Regulation minimum standards. The prominent conservatory is an essential feature in the low energy design.

This house achieved an equally low energy requirement figure with an exceptionally high level of insulation. The walls are of concrete poured into permanent polystyrene formers, with special insulating panels used for the roof and the floor. The technology is German.

This house at Energy World built by ASPP requires half the energy needed in the usual new home. It is not as energy efficient as the two houses above, but is wholly conventional in appearance and layout, with extra insulation and an advanced heating and heat recovery system. The right balance between energy efficiency and conventional appearance is very important when considering a new home as an investment.

ENERGY EFFICIENT HOUSE DESIGN
THE LLEWELLYN HOUSE

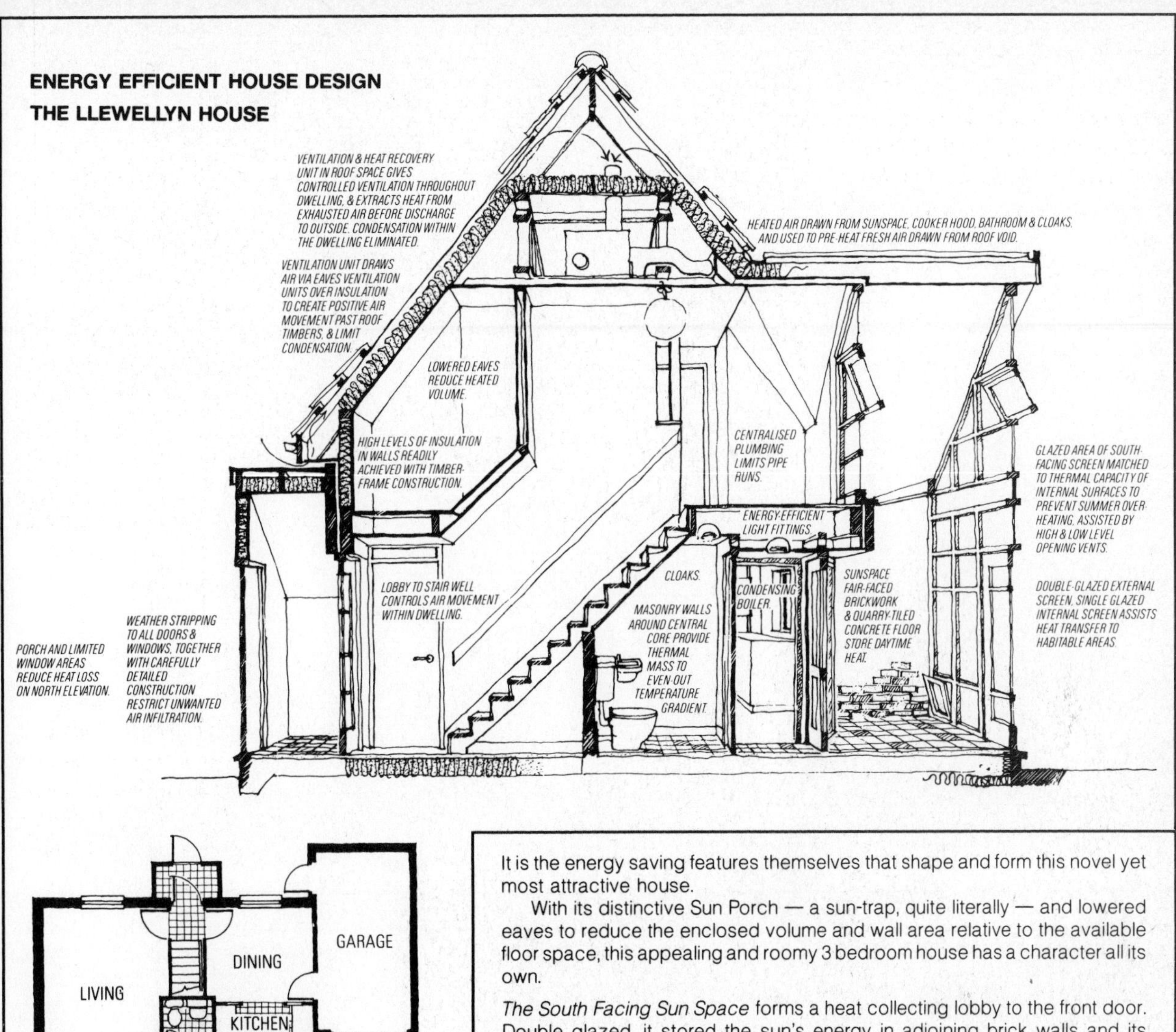

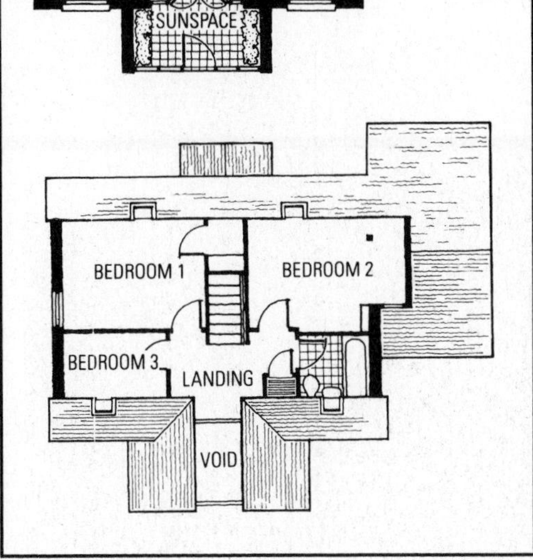

It is the energy saving features themselves that shape and form this novel yet most attractive house.

With its distinctive Sun Porch — a sun-trap, quite literally — and lowered eaves to reduce the enclosed volume and wall area relative to the available floor space, this appealing and roomy 3 bedroom house has a character all its own.

The South Facing Sun Space forms a heat collecting lobby to the front door. Double glazed, it stored the sun's energy in adjoining brick walls and its insulated concrete floor. This is then released during cooler parts of the day and evens out the extreme temperatures.

The Central Core has walls constructed in blockwork and acts as a heatstore. To restrict the movement of cold air within the building external doors are lobbied and the staircase void fully enclosed.

Warm air is circulated under the stairs by the ventilation system to maximise heat transfer.

A Heat Exchange and Ventilation System drawing warm air from the kitchen and bathroom, extracts heat to warm incoming fresh air including that from the Sun Space.

Stale air, robbed of its warmth, is then exhaled through vents in the roof. In this way, heat wastage is minimized, and condensation is controlled.

Central heating is by a high-efficiency wall mounted gas boiler and radiators.

House Insulation Standards — having collected and transferred our heat, it's vital that as little as possible is lost. By using additional insulation in walls and roof, 'U' values have been significantly improved. Walls are 0.216 as against the required 0.6; and the roof 0.26, as against 0.35.

Heat loss through the floor too, has been greatly reduced. Pre-stressed concrete beams with Polystyrene in-fill blocks give only 0.248.

Reproduced by courtesy of Llewellyn Homes Ltd, Eastbourne.

Details of the energy efficient design features of the Llewellyn Homes house at the Energy World exhibition at Milton Keynes. A photo of the house is on page 91, and it needs only one third the energy of a house of the same size built to the minimum standards required by the Building Regulations.

for insulation, not least in enabling you to build a slightly larger home with the standard level of insulation.

There are two areas where it is particularly difficult to get an accurate appraisal of the pros and cons of extra insulation, and this is with triple glazing and cavity wall insulation, both of which are advertised in a particularly aggressive way. Double glazing is now virtually mandatory as most houses have walls with over 12% of glazed area. Triple glazing only makes a significant contribution to heat conservation in modern construction when used for very large glazed areas. All sealed units are much the same, and increasing the width of the cavity has little overall effect on the heat loss figures for the whole dwelling, which is what we are concerned with. The best window insulation of all is provided by continental double windows with two separate casements, but they are not generally popular here. New types of insulating glass, like Pilkingtons' Kappafloat, give double glazing units the performance of triple glazing, and are likely to become widely used.

Advertising for cavity insulation overlooks the fact that the cavity was put into the wall for a purpose, which was to keep the weather out however badly a particular joint may have been pointed or however porous a soft brick may be. Early cavity insulation based on foams which filled the cavity had a reputation for causing damp patches. As a result, local authorities in many areas have regulations to control their use in exposed situations. Modern foam systems are more reliable, but the same level of insulation can be built into new walls while still keeping the cavity by using a slab system. This involves using special tie-irons in the wall to hold a slab of insulating material firmly against the inner skin, maintaining the cavity between it and the outer skin. This gives the best of both worlds.

All dwellings with a high level of insulation and effective draught proofing have a built in condensation problem. To avoid this the roof space must be adequately ventilated, extractor fans should be installed in kitchens, and sometimes bathrooms, and the occupants must learn to open windows if they are not accustomed to doing this. It is a chore, but it is the price of all this energy saving. The NHBC publish an excellent booklet on ventilation and condensation which all registered builders are supposed to give to their customers.

New and cheap ways of putting heat into a building are more involved, but come down to either being more efficient ways of using conventional fuels, alternative fuels, solar collection systems or heat pumps. All are now on the market and widely promoted.

Starting with better ways of using fossil fuels, there are whole ranges of new appliances of all sorts which have in common that they put more energy from the fuel into the house as heat, and less of it up the chimney. They are very much more expensive than older equipment, and like all complex gadgets they may have a shorter life. The rule is to see if they are approved by the relevant fuel advisory service such as the solid fuel advisory service and its equivalents.

Alternative solid fuel usually means logs or straw bales. Straw is usually available only to farmers, and boilers to burn it are a practical proposition. They are efficient, reliable, ugly, dirty and require an outbuilding of their own with a small yard to store the straw. The cost of burning logs depends on what they cost you, and how you view the work involved.

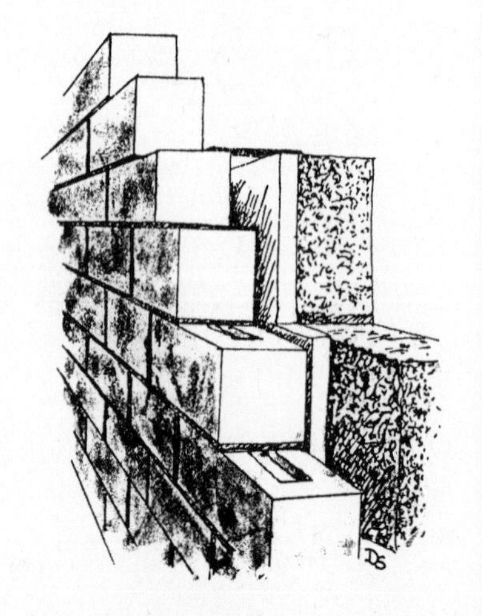

30mm insulating board between the two leaves of external walls can provide statutory insulation levels. It is fastened to the inner leaf to retain an air space between the insulation and the external masonry.

Finally, heat pumps. First of all, they are not new. Heat transfer technology has been with us since the 1890s, but what is new is that the price of oil for conventional heating has risen to a point where this technology can now be cost effective for domestic heating. Heat pumps are quite different from any other heating system in that they do not 'make heat' by burning anything, and instead collect heat from a convenient source and deliver it where it is wanted. This is not easy to understand but it helps if we realise that a domestic refrigerator is a heat pump working in reverse. It collects heat from inside the cabinet and dumps it to waste via the coil at the back. A heat pump collects heat from somewhere outside the house and pumps it into the house.

The source of heat can be a river, the heat in the ground, or the heat in the air. Almost all heat pumps take their heat from the air, and work however cold the air is. If the air is at 8 °C when it is sucked into the heat pump, it will leave it seconds later at 4 °C, and the heat extracted will end up as very hot water in the radiators in the house.

The appeal of the heat pump lies in the fact that it is more economical to use energy to collect heat and move it about, than it is simply to use energy to generate heat. Heat is a form of energy, and both are now measured in kilowatts. Using 1 kW of electricity in an electric fire generates 1 kW of heat, using the same kW of energy in a heat pump enables it to collect between 2 and 3 kW of heat and to shove it into radiators. This is a marvellous trick, but it has its price, which is the high capital cost of the machine.

Fifteen years ago an oil-fired central heating installation in a four bedroomed house cost less than 20% of its capital cost to run each year, but today this figure might easily be 75%. Running costs dominate all central heating economics, and a significant saving in running costs can justify a high capital cost.

From 1972 to 1982 oil prices roses at a rate which made the savings offered by a heat pump attractive in spite of the high capital cost, and a few thousand were installed in Britain, (and many tens of thousands on the continent). By and large, they have proved reliable, and gave the savings promised by the experts, although not the super-savings promised by some salesmen. However, oil prices have since fallen, and heat pumps have lost their appeal. When oil prices rise again they will be back. They were never really cost effective when the alternative fuel was gas or solid fuel, nor were they suited to a home of less than 2,000 sq. ft, but in the very long term they will probably be very important indeed.

If this chapter on energy saving appears somewhat cynical, this will serve to balance the over-enthusiasm with which the subject is usually approached. The more important a subject is, the more important it is that it should be approached realistically. Energy saving is very important, and the consideration given to it merits more than simple enthusiasm.

The author with his own heat pump. The wooden structure contains a heat store which holds heat collected using cheap tariff electricity at night, and which is used to heat the house during the day. The heat pump takes heat from the air.

Self-Build Faces

Case Histories

These case histories are based on information kindly given by the self-builders concerned, and the cost figures quoted are the real costs taken from VAT records kept to reclaim the tax. This ensures that no expenditure was missed out!

All the Associations appear under their own names, as do most of the individuals. A very few have pseudonyms to protect their privacy, but they are real people just the same.

Tididew

Tididew is an unusual name for a house but it explains itself as soon as it is read backwards, and is the home of the Sharp family who are seen in the picture at work on the garage which is the last stage in the new home. Their story is interesting because it demonstrates the sort of unexpected problems that can arise, and also how to get round them and finally to turn them to advantage.

Roy and Sheila Sharp's interest in building for themselves was first aroused by an article in *Practical Householder* in the late 1970s and they kept the magazine for several years, hoping that some day they would find a piece of land where they could put in to practice what they had read. They discussed the sort of house that they would build, and the day-dream just would not go away. Eventually, in 1983 they approached their local council and asked if plots were ever sold to self-builders. They were told that a scheme was just about to be advertised, and that as the first enquirers they could have the first set of tender forms. It all seemed too good to be true.

The plots were in a village on the outskirts of Derby, overlooking open country. The council documents gave all the details but there was no guide at all about how much they expected to be tendered. To get help with this the Sharps went to an Estate Agent, who suggested an average value for any of the plots of about £8000. Having read *Building Your Own Home* they then decided to approach Design and Materials Ltd. for further help. Geoff Gregory of the D & M field staff visited the site and went over all the probable building costs with them. After a great deal of thought they decided to tender £8100 for the largest plot, which was an unusual shape but which would suit the house that they wanted to build. The bid was accepted, and they were later told by the Housing Officer that the figure which they had put in was his own estimate of the plot value, which was a relief. Everyone buying by tender wonders if they have paid too much!

Planning Consent was considerably delayed by lengthy discussions concerning the actual position of the house on the site as there was a buried 11KV electricity cable running down one boundary, and there are strict rules about how near a building can be to high voltage cables. In due course everything was settled, the Sharps sold their house and moved into a caravan on the site, and work started. Almost at once the digger found a major problem: the cable was not where it should have been, but ran directly under where they intended to build the house. There was no way it could be avoided. When the electricity board officials visited them to look at it the problem was quickly found to be much more complicated than anyone imagined. The cable had been laid by the American army during the war and the casing was too brittle for it to be pulled about as a modern cable could have been. Worse still, somehow the easement for the cable was not in the title deeds, so that no one was responsible for it. The cost of moving it was astronomical. Everyone was very sorry, but no one could do anything and Sheila Sharp, with two small children in a caravan in the middle of a sea of mud, felt like going home to Mother. There were two things that could be done: to enlist a solicitor and make all the fuss that they could, or quietly try to persuade the Electricity Board to find an answer. They adopted the latter approach and won. The Board agreed to bend their rules and let them build on top of the cable, and then to lay a completely new cable at a later date at no expense to the Sharps at all. This enabled the house to be built exactly where Roy and Sheila had always wanted it, and work started again.

The Taylors at work on the garage to their new home which can be seen in the background.

The next problem arose when the first Building Society stage payment was due. This had been agreed as 20% of the valuation figure for the house, and as this was £55,000 the sum payable was £11,000. The letter from the Building Society specifically quoted '£11,000'. When the Society's surveyor arrived he approved the work and then casually mentioned that the letter was wrong and that the first payment would be 20% of the mortgage required, or £5000. He was very sorry but rules were rules. As the Sharps had paid for the land from the proceeds of the sale of their old house, the missing £6000 was essential to pay for their package of materials from D & M. Everything would grind to a halt unless it could be found. Obviously someone had made a very big mistake and there was a temptation to make a fuss about it; writing to Esther Rantzen was one idea! Instead they managed to get the Building Society to deal directly with D & M over the payments, and while the company and the society argued away the materials kept coming and the work kept going. In the end the Building Society released all of the money earlier than had been planned, so once again a problem had been turned to advantage.

There is a lesson for all self-builders in these two events: things will go wrong and sometimes it will not be your fault. If this happens your instinctive reaction will be to claim your rights and to make it clear that you are going to win. The result is usually a lengthy delay while everyone deals with the problem in the most formal way, and delays are very expensive. Roy and Sheila Sharp avoided this, and persuaded instead of insisting. It is a good example to follow.

Roy Sharp is a tanker driver and Sheila helps at a play school; neither of them had any relevant experience to help them with the building work, but they were determined to do as much as they could themselves. The 'they' included Sheila who did everything from digging foundations and drainage trenches to lifting the roof trusses on to the roof. The only sub-contractors employed were a bricklaying gang and a plasterer, and they did everything else themselves including the electrics, the central heating and the plumbing. They say they enjoyed every minute of it, although on reflection they have reservations about November 5th when a butane gas cylinder in the caravan caught fire and had to be put out with an extinguisher late at night, with the family standing outside in the mud, freezing cold, wondering where to spend the night.

The total cost, including the land was £38,000 but the VAT claim repaid them £3000 of this, giving a net figure of £35,000. The Building Society valuation before they started was £55,000 but the house is now worth a good deal more than this on the open market.

And it all started with an article in *Practical Householder*.

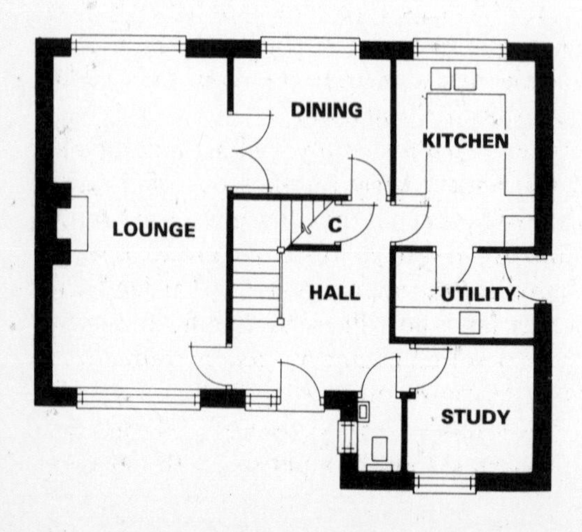

The Oakwood Association — The Good Average Scheme

When writing about Self-build Associations there is a tendency to seek out the interesting schemes, the Associations that are unusual in some way, have particular problems or are particularly successful. For many who are considering whether or not to join a self-build group this can be misleading, so this case history is about an average association, which started only five months before this was written and were photographed in their twelfth week of work. They are happily working on schedule, on budget, and are as average a group as can be found. The self-build scene revolves around them and a hundred others like them.

The Oakwood Association is a development of sixteen houses in a cul-de-sac off Bishop's Drive in Oakwood, three miles from Derby. It is in an area which is being developed very quickly, with eight well-known developers building three and four bedroom detached houses for commutors in a variety of cottage styles, all off the same road. This affords an interesting opportunity to compare what they all have on offer, and must be a stimulus to the Oakwood Association members who are building in the middle of it all.

The land was found and the scheme put together by consultants Wadsworth and Dring. Everything should have been ready for a start in April 1986, and after local advertising, two meetings were held in February to attract members. One took place in a local hotel, the other in the library, and both were exhibitions of photographs and the Wadsworth video, with people available to chat to visitors, without the speeches that used to make these occasions rather formal. Eleven people were signed up and all these original members are still with the group, having been joined by five others in the early stages of the work.

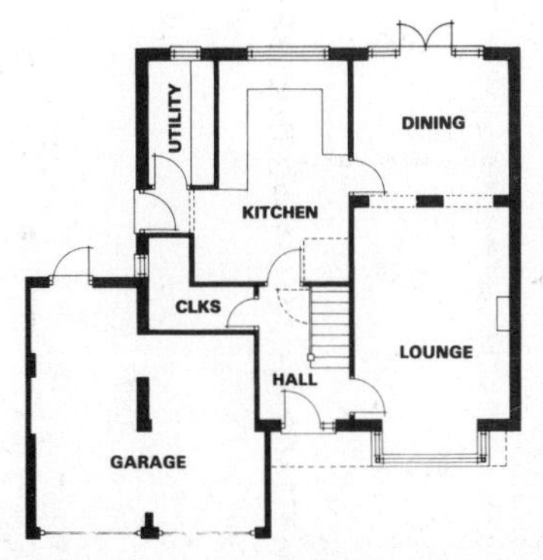

The sixteen units comprise a pair of three bedroom semis and a mix of three bedroom detached houses with single garages and four bed units with double garages. The budgeted construction costs run from £22,800 for the semi-detached units to £36,400 for the largest houses, and the provisional valuations for the completed homes are £30,000 and £48,000 respectively. Both building finance and the mortgages are from the Derbyshire Building Society. This is the first self-build association with which this Society has been involved and they are monitoring progress with keen interest. They are particularly concerned that their lending should not exceed the realisation value of the work on site, and in the early stages when most of the effort has been on foundations and services this is presenting a minor cash flow problem. It will be solved soon as the first four units are about to be tiled, which will give an appropriate boost to the valuation.

A feature of Oakwood is the very happy family feeling among the members; the enthusiasm and team spirit fairly bubbles out of the site caravan windows! Old self-builders will point out that they have yet to build through a winter, but making a start in the spring does help a group to settle down before the going gets tough. Among the members are two brickies, one plumber, and one industrial electrician who will easily handle domestic wiring, while the rest include three assorted engineers, one accountant, a printer, a physicist and a fitter. They also have a most unusual volunteer . . . a member's wife who puts in the same hours as everyone else, working as a carpenter. Pam Wright has always been a good DIY person, with her own tools and Black and Decker, and after a first visit to the site asked the foreman if she could work as a volunteer. In spite of having her own regular job, she puts in 25 hours a week at weekends and in the evenings, and is completely accepted as one of the team. 'At first they wanted to help me with anything that looked heavy' she said, 'but now they just leave me to get on with it.' A most unusual lady: self-build associations have never found a way to harness wife-power in the way that individual self-builders do, but Pam's contribution shows what the potential is.

The site foreman is bricklayer Mick Nichols, the secretary is printer John Newsham, while it was inevitable that accountant Paul Taylor should end up as treasurer. Group members are contracted to work 20 hours at weekends and 5 hours in the evenings during the week, and most are putting in extra voluntary hours which do not count in the accounting arrangements. The association intend to sub-contract some of the bricklaying and all of the double glazing to supply and fix contracts. Scaffolding, two mixers, a site caravan and a dumper have been bought at auction and will be sold on at the end of the scheme. The dumper has proved to be bloody minded, with a tendency to set off on trips of its own without any human involvement, but some machines have the wrong temperament to be good group members.

After 12 weeks the members have now got four homes up to square and foundations are in for the next stage. They have turned the individuals into a team and the race is now on to get as many units as possible roofed in before the winter. Costs are on target at a time when house values generally are rising. It is an average performance of a good average association, and as a bonus everyone is obviously enjoying themselves enormously. We wish them well.

Pam Wright working on the dormers of her own new house.

Above — the manager of the Derbyshire Building Society cuts the first sod in April 1986 with members looking on.

Below – twelve weeks later the first houses are being roofed. John Newsham, left, and Paul Taylor, right, with Terry Little, site manager for Wadsworth and Dring.

A Home in the Lake District

In 1985 Ian and Di Culley were teachers at schools in the Lake District with children approaching their teens and a house that felt too small. They knew all about self-building, had read *Building Your Own Home*, and for some time had been looking out for land on which to build. The intention was to find a plot in the £16,000 to £18,000 bracket. Eventually they found just what they were looking for . . . a third of an acre with outline planning consent on a steep slope with a view of the Old Man of Coniston on one side and the Crake Estuary and Morecambe Bay on the other. The price was £23,700. Fortunately their existing home had appreciated in value so the budget was revised and they bought it.

Ian Culley teaches physics and the theory of timber frame structures with their high thermal insulation and low thermal mass appealed to him. While waiting for the solicitor to complete the land purchase they collected a dozen brochures from timber frame companies, one of which had Prestoplan on the front cover. They turned it over, and staring at them from the back page were Graham and Pam Rhodes, two old friends from their college days who they had lost touch with years before. The Rhodes were standing proudly in front of their new Prestoplan home, so a telephone call to Prestoplan put them all in touch. In the euphoria of it all the Culleys were quickly signed up by Nigel Heathershaw, Prestoplan's regional manager. They have never regretted it for a moment.

Prestoplan suggested that they engage a local architect to handle the design of a special home to suit their steeply sloping site, and introduced them to a practice in Windermere where detailed plans were drawn from Ian's own preliminary sketches. After the usual discussions and alterations, followed by a quotation from Prestoplan, the planning and building regulation applications were made in December 1985.

Meanwhile the solicitor had hit a snag. The plot was sold with an existing main drain connection, and with a covenant in the title requiring that the next door neighbour should 'approve the design and siting of the dwelling, such approval not to be unreasonably withheld.' Unfortunately the neighbour did not want the new bungalow sited where it could use the drain connection, and instead wanted it lower down the slope. This could have led to an impasse if the covenant had not used the phrase 'not to be unreasonably withheld.' This wording means that in the event of a dispute the question of what is, or is not, reasonable can be resolved in a court. This is a great stimulus to negotiation! With both the Culleys and their new neighbours being sensible people, it was all settled amicably, with a new drain run into a manhole in the neighbours' garden.

Everything was settled in time for a start at Easter 1986, when a local builder was engaged to make an access into the site and to construct the involved foundations which are the price of building on a hillside. When this was done the Prestoplan erection team arrived and put up the timber frame in just eight days. The tilers followed on immediately and in less than a fortnight Ian and Di had a house that was waterproof, could be locked up, and was theirs to finish off by their own efforts. With the help of Joe Jaggar, a family friend who is a retired electrician, they handled all the joinery, the plumbing, the electrics, the installation of sophisticated heat recovery system . . . and the boarding.

Fixing the boards for the internal walls of a timber frame home is called 'dry lining' and involves very special techniques which give the same

Above – Ian Culley in front of his house

Right — and busy at his dry lining.

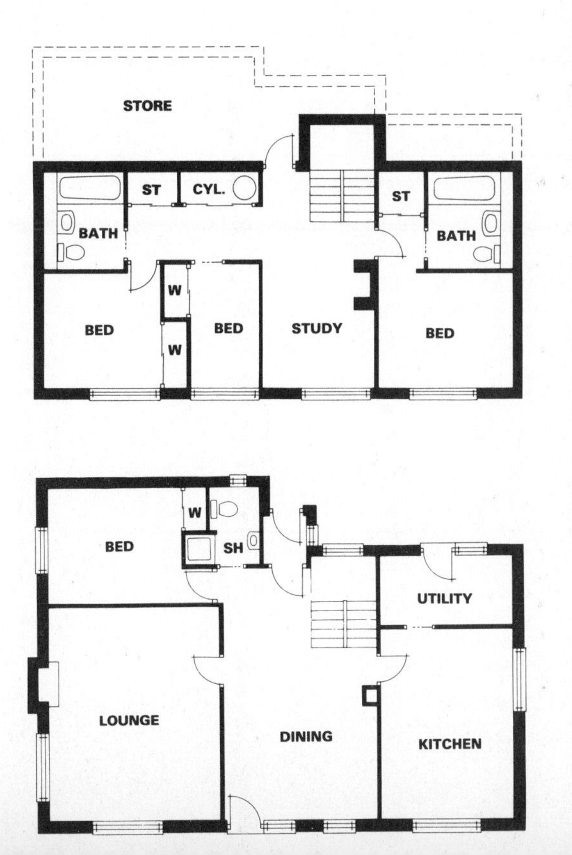

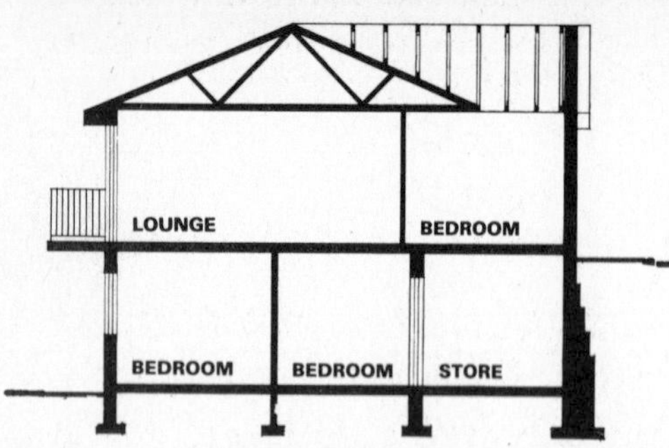

A rear view of the house with a cross section of the building.

standard of finish as classical wet plastering. To learn all about this Ian got himself a place at British Gypsum's plasterers' school at Carlisle, where he spent a week receiving a very detailed training. British Gypsum often helps self-builders in this way, although it seems unlikely that their profit on a £1,000 worth of plaster board makes the cost of the training worthwhile. It certainly gets them lots of good will. £1000 worth of plaster board weighs 7 tons, and Di remembers that she carried most of it down the site and into the site herself. Although British Gypsum had lent Ian a board jack, fixing 13 mm gyproc boards is a team effort and every one was a triumph of family togetherness.

When the photographs were taken in October the Culleys were only waiting for a special stair ballustrade to be delivered and fixed before moving in. The total expenditure, including the cost of the land and before making the VAT claim, is going to be about £70,000. The minimum market valuation is £90,000 and it is the house the family have always dreamed about.

The heating arrangements are interesting. Ian teaches physics and calculates that the heat loss from his new home will be only 3.6 kw. The family like open fires, so to meet the very low heat input requirement there are two fireplaces, one of them on the lower level with a stack running up through the centre of the house. Back up heating is provided by strategically positioned night storage heaters, and a sophisticated Husqvana heat recovery and ventilation system has been installed in the roof.

Zenzele — a Scheme with a Purpose

The Zenzele Housing Association in Bristol is quite unlike any other described in this book, and was set up as an experiment in breaking the 'no job experience — no job opportunity' trap in the run-down St. Paul's area. It was originally called the Bristol Pilot Project for the Unemployed and the intention was then defined as 'to provide incentives and work experience for the young unemployed so that skills acquired will improve their job prospects'. The way in which this was approached was to set up a self-build housing association for twelve members described in the press as young, disadvantaged and unemployed, and this was made possible by some forceful management by a local J.P., Mrs. Stella Clark, and a very effective management team which she assembled.

Their first coup was to get the agreement of the D.H.S.S. that association members who were still unemployed when they finished building should have the interest on their mortgages paid by the department as if they had incurred the mortgage before they became unemployed. At that time the ceiling payment of this sort was £22.50 per week, which facilitated a mortgage of £10,800. The Bristol and West Building Society was persuaded to offer such mortgages, and the NatWest Bank agreed that the association could have an unsecured overdraft. Building finance was arranged through the Housing Corporation. All that remained was to see if a scheme could be put together for new homes to be built for £10,800.

Zenzele members Joe and Chris Gordon and Colin Simpson with Mrs Stella Clarke, JP, outside the flats built by the association (photo: R.I.B.A.)

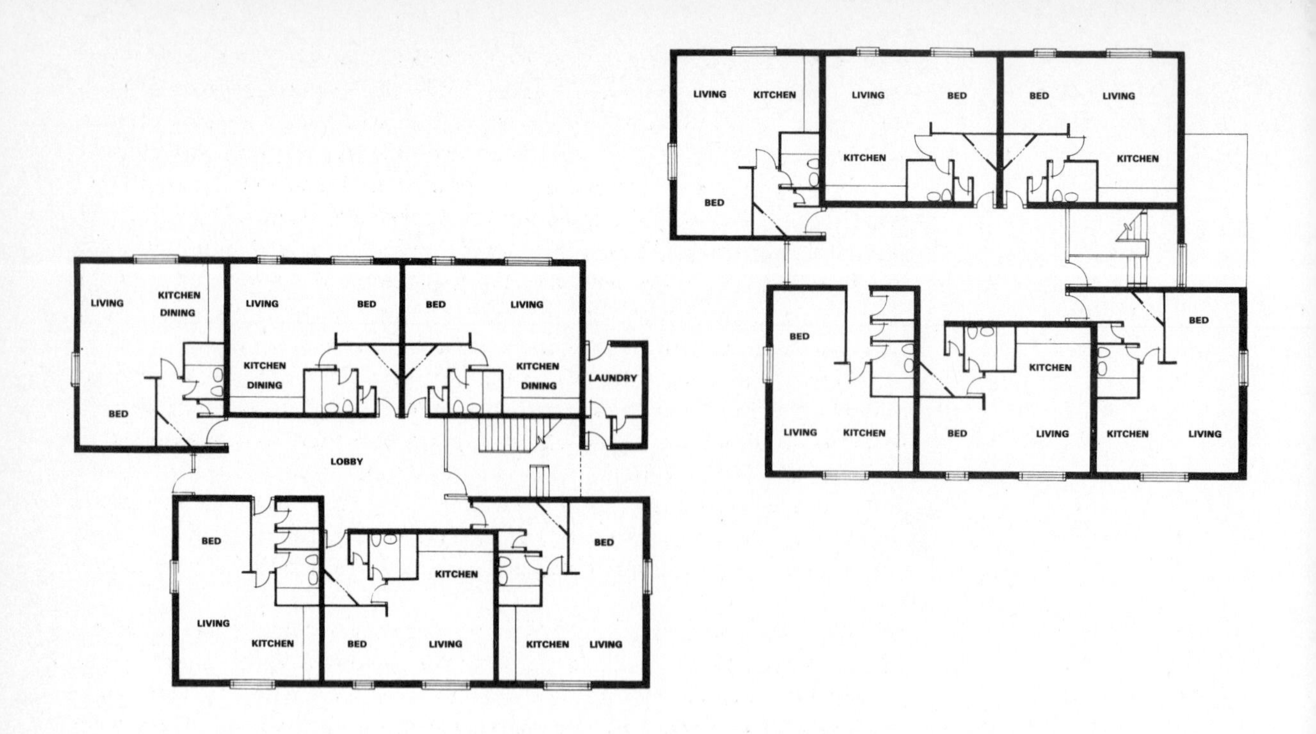

Prince Charles talking to Anthony Edwards, the job architect for Atkins and Walters, at the opening of the Zenzele flats.

A site was found on the open market at £15,000, and architects Atkins and Walters designed a simple and cost effective block of one-bedroomed flats that could be built within the budget. The next stage was to form an association, and instead of the usual advertisements and public meeting the first four prospective members were assembled by Tana Adebiyo, another of Mrs Clark's team, who was a community worker at St. Paul's. Their first task was to find another eight members, all of them unemployed, although some had building industry experience. The association was registered with the name Zenzele, which is Zulu for 'together'. The Federation's model rules were adopted and in late 1984 the association was in business.

Progress was at first quite fast, as members were on site for a full working day. The budget provided for the services of a working site foreman, and the first one appointed was a groundworks specialist as there were site problems with the cellars of old buildings. This involved some anxious moments as this sort of situation can never be costed with absolute certainty, but after various worries everything worked out. The groundworks foreman handed over to another who was to take the building on to completion, and the walls started to rise.

Construction was entirely traditional, and the members handled all the trades themselves. The only contractors brought in were specialists to lay the concrete slab first floor, which, together with a 3″ screed, had been designed to give the maximum sound insulation between individual flats. There were no building problems, but one very happy organizational problem . . . members started to get jobs. The whole scheme was based upon the idea that people find it easier to get employment when they are already working and this is just what happened. The members were soon in two groups: those who were still unemployed and could put in a normal working day and those with jobs who worked on the flats in the evenings and at weekends like most other self-builders. At the end of the scheme eleven of the twelve members had found jobs, which was the original intention behind the whole project.

The block was completed at the end of 1985 by which time costs had escalated by 11%. However the Building Society valuation of the individual flats had gone up to £18,500, and rises in the D.H.S.S. mortgage interest ceiling made mortgages of £12,000 possible. This not only covered building costs but also cookers, carpets and full interior decorations.

In April 1986 H.R.H. Prince Charles visited Zenzele and met the members. As a result the scheme got a great deal of publicity and it is to be hoped that it will be the first of many others of the same sort. However, an undertaking of this sort that failed would be particularly unfortunate, and management of the calibre of the management of Zenzele is essential if similar projects are to get off the ground. This is a challenge to the self-build consultancy moguls, and as this book goes to print we understand that it has been accepted. Two similar schemes are in the project planning stage, and one at least is nicknamed 'Son of Zenzele'.

Bungalow in Northants

Neil and Laura St John built a bungalow in Northamptonshire in 1985 as a result of reading the sixth edition of *Building Your Own Home*, and wrote offering their cost analysis as a case history. Neil is in the timber business, Laura has her own word processing agency, and presenting data comes easily to them as seen from the enclosures to the letter below. This arrived in May 1986; the photographs were taken in August '86. Between them the photos and the figures say it all.

Mr M Armor
c/o Prism Press
2 South Street
Bridport
Dorset

22 May 1986

Dear Mr Armor

Having read your "1984/5 Building Your Own Home" book I was enthused to find a plot and build my own home. I successfully convinced my wife it was a good idea and we promptly put our property on the market. We located a good sized plot with detailed consent for a small bungalow and garage.

We felt the plot could well justify a larger property and made arrangements to meet a planning officer on site to discuss the possibilities. This was well worthwhile and after discussion our proposed plans were slightly amended and submitted for approval.

Plan Sales Services provided the drawings and I must say we were extremely pleased with the overall service. By the time our house was sold planning permission had been given and contracts exchanged on the purchase of the plot. The very same day of moving into our rented accommodation our JCB was on site levelling and digging foundations.

I have enclosed detailed costs and other information which I felt might be of interest to you and wonder if you might consider covering our case/story in your next edition.

I am able to provide photographs of work completed at various stages showing the progression from foundations to topping out and the finished result. Needless to say if you require any further background information I could provide this without delay.

I look forward to hearing form you.

Yours sincerely

Neil St.John

Neil and Laura St. John in the garden of their new home.

Costings for the traditional construction of an individually designed four bedroomed detached bungalow with double detached garage

---

Plot size 0.26 acres

Bungalow floor area 1500 square feet Garage floor area 350 square feet

All works were sub-contracted to locally based craftsmen. Only co-ordination of labour and decorating carried out by owners.

The property has been built to a high standard and offers excellent insulation characteristics to include floor insulation, high U value insulating blocks, foil backed ceiling plaster board, high performance windows all with 20mm sealed double glazed units. Included within the finished price are solid Pine internal doors, built-in wardrobes to three bedrooms, fully fitted German kitchen and Jetmaster open fire.

<u>Programme</u>

Plot located with outline planning permission/decision made to go ahead/property on the market	November '84
Detailed planning permission approved	March '85
Completion of sale of property/move into rented accommodation/ commencement of works	April '85
Completion of works/occupation of new property	November '85

<u>Costs</u>

Purchase price of plot	£15,500
Net building cost after refund of VAT	£38,000
Total	£53,500
Minimum valuation by Bank surveyor for rebuilding insurance purposes, November '85	£78,000
Estate Agents market valuation, December '85	£89,000

Quotations received from local firms to carry out all building work ranged from £44,250 - £56,500. The bungalow was completed to a much higher standard than the builders specifications allowed.

Detailed costs:-	Cost (£)
Planning, Legal & Technical	
Plans	286
Legal fees purchase of plot	178
Planning consent	47
Building regulations	117
Contractors all risks Insurance	202
Architects certificate	280
Plant & Equipment Hire	
Cement mixer	50
Dumper	112
Vibrating plate	22
Petrol & Diesel	25
JCB	457
Brick cutter	16
Cowley level	6
Scaffolding	400
Sub-soil removal	150
Truck hire	71
Plant Purchase	
Cement mixer	188
Services	
Water connection	113
Water pipe	19
Accidental damage to electricity mains cable	72

Labour & Materials	Cost (£)
Bricklayers - To include:- Setting out, lay strip foundations, concrete blocks to DPC, concrete oversite, build up to wallplate & gables, lay path all round bungalow, all drainage, drive.	6837
Ready mixed concrete - strip foundations & oversite	1200
Hardcore	390
DPC/membrane	54
Sand & ballast	394
Cement	500
Lime/additives	17
Concrete blocks	436
Insulating blocks	774
Facing bricks	2352
Commons	91
Windows & external doors	1659
Garage doors	216
Lintels	470

Detailed costs, continued...	Cost (£)
Wall ties & joist straps	66
Jetmaster open fire	411
Chimney liners/pot	106
Wallplate/roof binders/ ventilated soffit & fascia timber	413
Roof trusses	929
Carpenter - trusses, fascia & soffit	400
Roof tiles - supply & fix	1789
Labour & Materials	
Drainage - all pipes, inspection chambers & pea gravel	698
Glazing - supply & fix	421
Ceiling plaster board	240
Labour - fix ceiling boards	200
Plaster	365
Plastering - labour	900
Floor insulation	156
Labour - fit floor insulation, lay floor screed	472
Internal door linings, architraves, skirtings, window boards	473
Internal doors	859
Carpenter 1st & 2nd fix	715
Nails, sundry timber	82
Ironmongery	185
Electrician - supply & fix	960
Plumber - supply & fix all rainwater goods & oil fired central heating system	3008
Sanitary ware (Fixing included in plumbers fee)	1161
Kitchen - all units and hob, extractor, double oven, sink with monoblock tap	2750
Carpenter - fit kitchen	180
Coving	32
Labour - fit coving	50
Artex - supply & fix	230

Labour & Materials cont.	Cost (£)
Floor & wall tiles	346
Tiler	300
Paint & equipment	348
Loft insulation	220
Patio slabs	199
Shuttering for paths	27
Drive kerbing	51
Drive gravel	57
Total	£38,000

Small is Beautiful — Associations of five or six can work very well

Withers and Co. of Northampton are management consultants who have recently formed three self-build Associations that are much smaller than usual . . . two of five members and one of six. One of them, the Rushden Self-Build Housing Association has just completed its five houses and the members are moving in to their new homes, while the others at Daventry and Rothwell are busily building.

The Federation of Housing Associations firmly recommends that associations should have at least a dozen or so members, and Andrew Withers, founder and managing director of Withers and Co. gets round this by enrolling both husbands and wives as members of these small associations. This seems unusual, but it is well within the rules and at any rate, Withers and Co. certainly seem able to make it work. It means that they are able to develop smaller sites which would not be large enough for associations of the usual size.

The Rushden scheme has the advantage of an exceptionally attractive site. It is within walking distance of the town centre, at the end of a cul-de-sac with a school opposite, a modern old persons' home to one side and open fields

Andrew Withers of Withers and Co. at the Rushden site of five houses.

Photo taken by the Nationwide Building Society for their magazine 'New Housing', showing the five members of the Rushden Housing Association on their site before starting work. The other two in the picture are Nationwide manager Robin Bailey (left) and Andrew Withers (fifth from right).

behind. The whole area is very well wooded and there are mature oak trees on the site. The five houses are all four bedroom detached units with either single or double garages, and construction started in September 1985 and was completed in ten months. Everything worked as with a full size association, except that there were no formal committee meetings as everything could be settled over a cup of tea in the site hut. There was a chairman, a secretary and a treasurer, but with only two other members these positions were pretty nominal!

The architect for all the Withers and Co. schemes is Trevor Jolley R.I.B.A. of Roade, and Ian Geddes R.I.C.S. of Northampton is the Quantity Surveyor. A computer print-out headed 'Shopping List' which was prepared for the association was particularly impressive, and set out all the materials required in a way that seemed far more convenient than the usual bill of quantities. The programme for this was written by Ian Geddes himself, and is used for all the houses with which he is concerned. Computer generated data like this is becoming generally available, and is particularly useful for the self-builder provided that it is written in plain English and not in professional jargon.

The estimated cost of one of the Rushden houses in 1985 was £39,500 including the land at £9,200, and the final cost in August 1986 was £41,250. The building society valuation was £66,000 and the open market value is currently £72,000. Finance was provided by the Nationwide Building Society, who also offered mortgages, and the whole project demonstrates how well a mini-association can work.

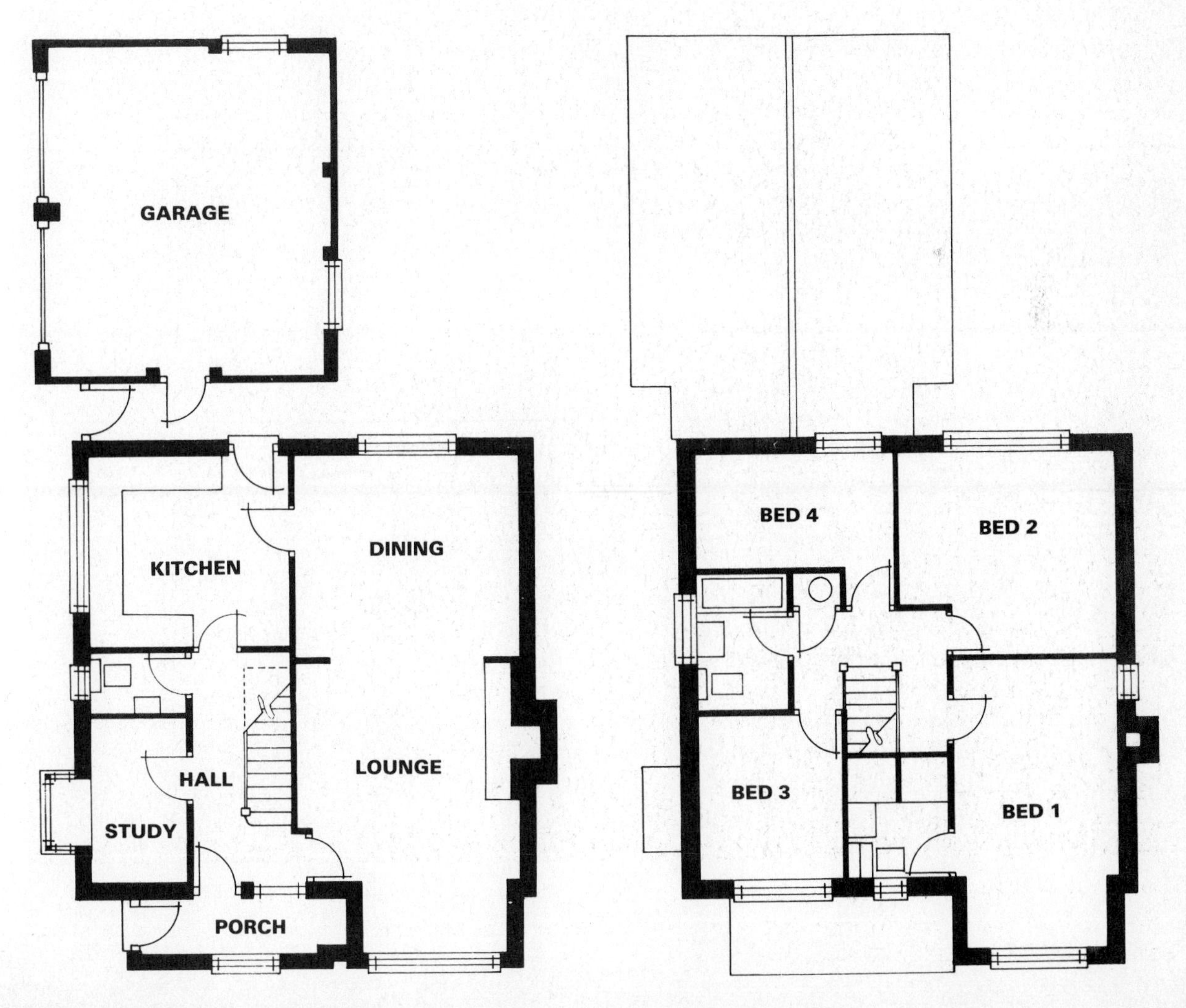

A Timber-frame Bungalow at the Top of the Market

One type of self-builder meriting special mention among our case histories is the individual who has had a long and successful career as an entrepreneur and who decides to round it off by building a new home before he retires. As his family will have left home the new house or bungalow may be smaller than he is used to, but the standard to which it will be built, and above all the site, are most important to him.

Our new home in this category is a timber frame bungalow built in just thirteen weeks by Squadron Leader Cowling in a tiny village in the West Midlands in 1986. Ted Cowling has had careers in the RAF, as a farmer, and in half a dozen businesses. With his children grown up and well established on their own he planned to sell his large house as a nursing home and to build a new bungalow on part of a ten acre agricultural holding which he had had for twenty years. In 1986 he realised this ambition after five years of planning battles and two public appeals. A brief resume of these legal struggles is interesting.

The Cowlings could not seek a conditional planning consent as a family 'wholly engaged in agriculture', and had to make ordinary applications, knowing that they would get local support, and expecting that consideration would be given to their long association with the village. Such applications were made on a succession of sites over four years, and all were turned down. A well known firm of Planning Consultants were retained to enter an appeal, and as there was local support they recommended that this should be a public appeal. It was lost. Ted Cowling persevered. After the statutory interval of a full year the matter was raised again, and this time everything was arranged with great care to give the best chance of winning. Twenty five separate letters of support were obtained from virtually every resident in the village. At an earlier appeal the council had successfully contended that there were opportunities in a neighbouring village for someone with local interests to buy building land, so the owners of every single scrap of vacant land in the village concerned were approached to sell, and all confirmed that their land was not for sale at any price. Finally, a drive was constructed into the site of the proposed bungalow, the plot was fenced with attractive wooden fencing, trees were planted, and a garden laid out; all to show what was intended and to demonstrate the high standard to which the proposed development would be carried out. A Q.C. was engaged for the appeal hearing — and consent was obtained.

Now whether you give three cheers, or whether you think this demonstrates that there is one planning law for those concerned to have their own way and another for the less determined, there is no doubt that the appellant showed his intention of building a new home that would be a real contribution to the village, and this is what the planning acts are supposed to be all about. It must also be pointed out that key factors in the success of the appeal were the total local support, and consideration of the family involvement in the local scene. An equally strong and determined figure from outside the area would not have had the same success.

With the outline planning consent tied up, the next consideration was the detailed design of the building. Squadron Leader Cowling's son Stephen, who is a civil engineer who had built his own home, suggested to his father that he should read *Home Plans for the Eighties*. This was obtained from the local

library, and promptly had its cover eaten by the family Golden Retriever which meant that the library had to be paid for it! In spite of this poor start the book was studied with care, and the Prestoplan 'Richmond' design, which was one of 230 designs illustrated in the book, was chosen as being well suited to the site if it could be modified in a few respects. Prestoplan were asked to send someone over to discuss their service, and the man who arrived was John Ellis, who projected such a 'can do' image that he quickly got the job. As if to atone for their earlier hostility over the outline consent, the local authority quickly passed the detailed plans.

A start was made in the spring, with Stephen Cowling setting out the site. A contract was placed with a local builder to supply the labour to put in the foundations and drains, to power-float a floor slab, and then to build the external brick walls to clad the timber-frame structure. All the materials were purchased by Ted Cowling, who made full use of his local contacts for this, both to get keen prices and to make sure that everything was of the very best. On March 16th, two weeks after starting, a Prestoplan erection gang came with a caravan in which they lived while they were putting up the frame. They worked from 7 am to 9 pm and completed the job in a week. Once the frame was roofed the various sub-contract tradesmen started work, and on June 6th all was finished, thirteen weeks after starting. The lawn was turfed, trees planted, and everything looks as had been described so often during the long drawn out planning process.

An interesting feature of the house is the continental type direct hot water system with mains pressure on all taps and instantaneous water heating. This does away with both water storage tanks and hot water cylinders, and means that fuel is only used to heat water that is actually required. The system installed is oil fired, although similar instantaneous systems are available for either gas or electricity.

The value of the bungalow is right at the very top of the market, with the lovely position helping a lot with this. The costs reflect many special situations and are not typical, but of all the good investments made by its owner over the years, this new home must be one of the best.

Doing it all on your own

Alan Brooks is a social worker and his wife Gwyn is an English teacher and until they set out to build their own home in 1985 neither of them had ever had anything to do with building at all. 18 months later they moved into a very attractive bungalow built to a meticulous standard, and they had literally built it themselves with no professional help at all except for a plasterer and a plumber. On top of full time jobs and the demands of a teenage family they did it all the hard way, digging foundations by hand, mixing concrete instead of using truck mix, laying bricks, tiling the roof, and learning enough about electrical wiring to handle their own electrics. They knew that they were taking on more than is usual, but to them the important thing was literally to do it themselves and live in a home which would really be all their own work. It is easy to see in this a social worker's reaction to a demanding career without an end product, and a deep urge to do one job at least that would have a tangible result. Leaving such philosophising aside, their home is a textbook example of all that they had hoped for and is a remarkable achievement.

They story started in February, 1984 when *Building Your Own Home* was an alternative selection of the month for members of the World Books Book Club. The Brooks were members, and when the book arrived the idea of self-building took hold at once. They live in a small village near Lincoln which is a part of England where it is still possible to find a building plot in a village street at a reasonable figure. By a remarkable coincidence a building plot was put on the market in the village at just the time that they were reading the book and they decided to take the plunge.

Their first step was to telephone D & M Ltd., which led to one of the company's field staff coming to advise on what was practical and to discuss budget figures. Their ambitions seemed realisable and the contract for the plot was signed at a figure of £14,750. The only problems were those presented by the mature trees on the site which both the planners and the Brooks wished to retain as far as was practical, and which Alan was careful to take into consideration in a design which he drew himself. D & M produced the detailed drawings, and the usual dialogue with the planners began. While this was going on the family sold their existing house and moved into temporary accommodation only a stone's throw from the site.

On December 1st, 1984, they made a start, clearing the topsoil by hand and digging the foundations with two spades as they did not want a JCB on the site for fear of damaging the trees. Two spades, because Gwyn played a full part in all of this right from the beginning. Concrete was mixed in a secondhand mixer and barrowed into the footings. Bricks, blocks and all other materials had to be unloaded at the site entrance and carried in by hand or in a wheelbarrow. It was slow work, but compared with some other self-builders they were in no hurry. This was a good thing, as they had to learn everything as they went along. They are a couple who are meticulous in everything that they do, and they were determined that all the work in the new house would be to the highest standards. The brickwork can be seen in the photograph and any professional builder would be proud of the detailing of the eaves and the corbells.

Be September 1985 the roof was on, with the help of a friend who was recruited to help them lift up the roof trusses. The normal D & M contract includes supply and fix roof tiling, but the Brooks would have none of this: they would tile their own roof. Gwyn carried them up the ladder and Alan laid them. It is one of the best tiled roofs for miles around.

With the roof on, the work could progress under cover and the family finally moved in during April 1986 with much still to do. The total cost had been £38,164 including the land, and was reduced by £2,575 reclaimed from the VAT authorities. This gave a total expenditure of £35,589 which was right on the original budget. The valuation made when the house was first occupied was £55,000, and local property values have risen since.

The work was financed by bank stage payments which were consolidated into a bank mortgage when all was finished. It is interesting that this arrangement was chosen because the bank offered to be more flexible over cash flow requirements than the Building Society. As often happens, the bank manager got so involved that he became a regular visitor to the site to see how they were getting on . . . in a queue with the Building Inspector, the planners, the services engineers, the certifying architect and all the other official visitors that pester self-builders.

It must be emphasised that this amount of physical work put in by the Brooks family is more usual among group self-builders, and to this extent this is an unusual case history. It is also a super story of interesting people who got total satisfaction from literally doing it all themselves.

An Airline Pilot's Home in Gloucestershire

This is a case history about a self builder who was able to make use of a special situation to build on a piece of land which he had had his eye on for years. This gave him a flying start, as will be clear later, and it is interesting to find how many self builders acquire their land in an equally complicated way.

Chris Jackson is an airline pilot, and his wife Sue is a well known breeder of show dogs. They lived in a very attractive Gloucestershire village for sixteen years, and at the side of their garden was a paddock with a half built bungalow that had been partially tiled and then abandoned. For complicated reasons it did not have any access at all, which is why it had been abandoned, although there was a way in if the Jacksons sacrificed their front garden. They had no intention of doing this unless they could buy the site, build a new home on it, and then sell off the old home without a front garden. In 1984, after nearly fifteen years of this situation, they managed to buy the paddock for £16,500.

The planning situation was interesting, as the planning consent for an uncompleted building does not expire. In theory they were able to carry on with the bungalow, but of course the Jacksons did not want to do this. They met the planning officer on site, fairly confident that they would be able to negotiate to build the house that they wanted. To their delight they were told that if the existing structure was demolished they could have consent for two new houses. The paper work for this came through in early 1985, and they immediately sold off one plot for more than they had paid for the whole site. Of course, this notional profit had to be set against the loss in the value of their existing home when it lost its front garden.

In the meanwhile Chris Jackson was looking at all his options in building, and as pilots have plenty of reading time while waiting at airports, he was able to look at all the alternatives very carefully. At an early stage he read Building Your Own Home, and following this he sent for literature from three package companies, — Potton, ASPP, and D & M. He looked carefully at the service, costs and designs offered by all of them, and decided that the D & M Chatsworth design was what he wanted. However, he was concerned at being so far from the D & M head office, and felt that he should look into using an

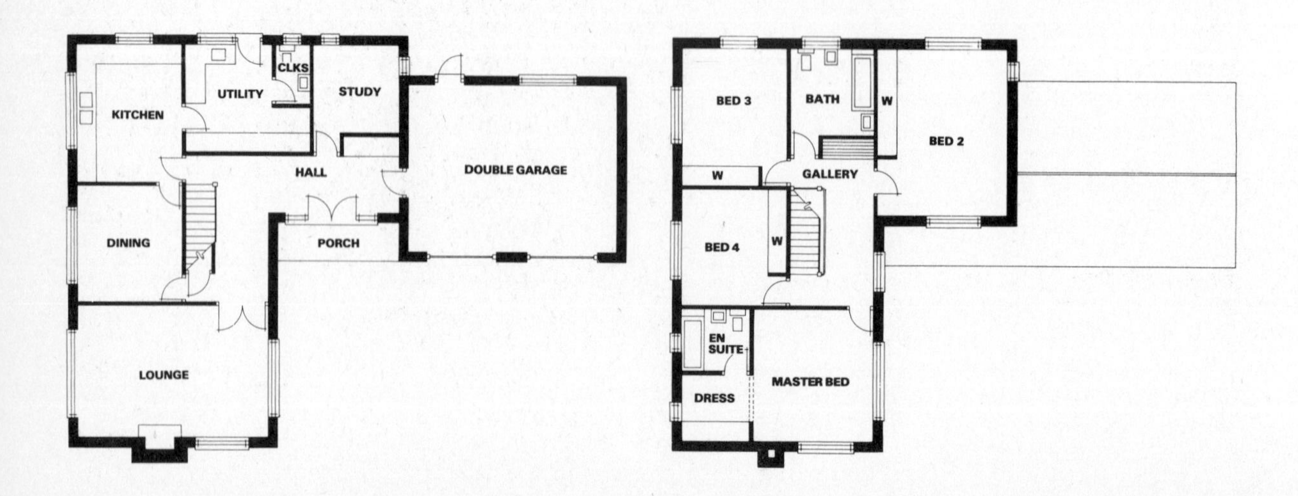

Chris and Sue Jackson at their home in Gloucestershire

architect instead. On his rest days between flying he invited three different architects to meet him on site, and showed the D & M proposals to each of them with the question 'can you design me something better for this site and have it built more cheaply?'. None of them persuaded him that they could, so he signed a contract with D & M. Concern over liaison and trouble shooting was eased after meeting David Snell, a larger than life extrovert who lived in the area and has ten years experience of looking after private clients for D & M.

Planning consent was obtained on 28 March 1985, and work started on site shortly afterwards. Construction was in Bradstone, and the Jacksons were fortunate in obtaining the services of the bricklayers who built the Bradstone show bungalow at the Building Exhibition at the National Exhibition Centre, and the work is certainly to exhibition standard.

The whole job was carried through without any serious problems, and during it they had another marvellous stroke of luck. Just after the new house was roofed the estate agents' board was delivered to advertise their old home for sale. While it was being unloaded a car stopped, and an offer was made and accepted on the same day. The board was never even put up. And it couldn't have happened to nicer people.

The D & M package of materials, with Bradstone walling, Bradstone stone slates, and mahogany joinery throughout, cost £25,545, and the total cost, including the D & M materials and with a very high standard of fixtures and fittings, was £66,406. This includes carpeting, phones, etc — every cost involved in the move, and not just the bare building figures. The value of the new home has been variously estimated at between £100,000 and 120,000.

The Jacksons cleared the site themselves, demolished the old building, did all the painting, gave a most professional finish to all the mahogany, and endlessly moved materials about to keep the site tidy. Sub-contractors were employed for all the trades, and the family seems to have made great friends of all of them. When flying as a passenger in a jumbo jet one tends to wonder if the pilot is up to his job. If management performance on a building site is anything to go by, one can feel reassured.

Suffolk Gardens Self Build Association

The Suffolk Gardens Self Build Housing Association at Marsden near South Shields is important because it is an example of urban renewal, which is both trendy and very worthwhile. The site was formerly covered with concrete panel prefabs which were very run down. These were demolished by the Borough of South Tyneside and the site made available for self build housing at the very modest price of £75,000. It is very important that selfbuilders should get their fair share (or more) of land made available by urban renewal schemes, and the more publicity that this scheme gets the better it will be for selfbuild as a whole.

This drawing was prepared for the front of the brochure giving details of the sceheme which was presented at the public meeting in November 1984

The Association is managed by Wadsworth and Heath of Cleckheaton, and has moved very quickly since it was launched. Everything started with an advertisement in the local press in the Autumn of 1984, following which there was the usual public meeting in the Town Hall. One of the speakers was the Chairman of the Housing Committee, as the Borough has been concerned with two other successful self build schemes in recent years, and the council gives enthusiastic support. The proposals put forward were for an Association to build 24 houses at a projected total cost of £530,136, against a projected value at 1984 prices of £726,000. All that the members were asked to contribute was a lump sum of £250, the capacity to repay a mortgage in due course, and eighteen months of 28 hours of work a week. A very strong Association was quickly formed with a preponderance of building industry tradesmen.

The start on site was delayed until March 1985 due to legal difficulties over a road closure and a footpath diversion, but once the Association got going it has forged ahead in a very dramatic way. By late September 1985 when this feature was written, eight houses were completed and ten more were under construction. Costs were within the budget, and the market value of the completed homes is above the 1984 estimate.

There are two house types, six pairs of semis costed at £20,081 per unit and twelve detached houses at £24,098 each. This includes the land and services.

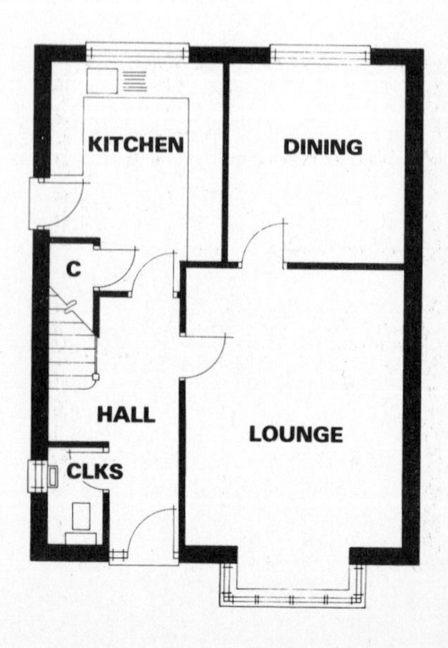

Floodlit! Working through the night

The 50:50 mix of semi and detached houses was laid down by the local authority although the members would have liked a higher proportion of detached units. At the original meeting the managers put forward estimated valuations of the finished properties at 1984 prices prepared by a leading firm of local estate agents. These were £27,500 for the semis and £33,000 for the detached houses, giving a margin on costs of 27%. In September 1985 the committee felt that current figures were £35,000 and £42,000, giving a margin of 43%.

In the first six months of the scheme the members worked 28 hours a week, but anticipate reducing this to 26 hours a week as the evenings close in. They attribute their success to tight discipline, and to the pace at which they have kept things moving. Everyone says that the managers were essential to the project, but some feel that the group is too large. Only two members dropped out in the six months before the work started on site, and none have since. It is an association of hard, competent men with a no-nonsense approach — typical of the North-East, and a good example of an urban renewal scheme of the sort that is wanted in so many of our industrial areas.

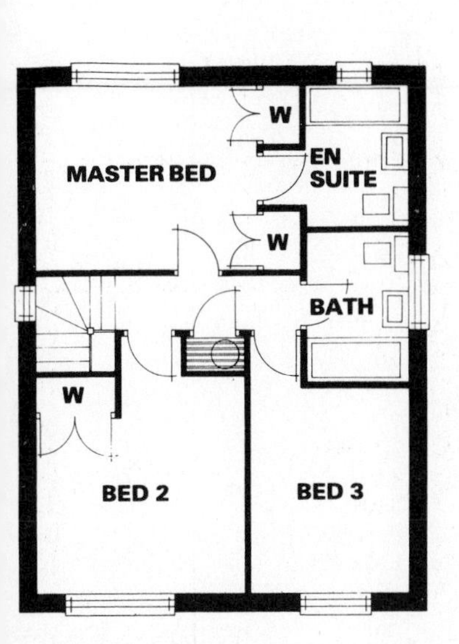

Topping out

A Prestoplan Home in Wiltshire

The letter on this page is dated 22 October 1984, and provided a splendid opportunity to keep in touch with a typical self-build job that, at that time, had not started. It proved to be an exceptionally fast moving project, and although various set-backs prevented work on site starting until January 17th 1985, the Case family moved into their new home on April 1st. Needless to say this speed of construction involved building with a timber frame, and a great deal of careful organisation and planning.

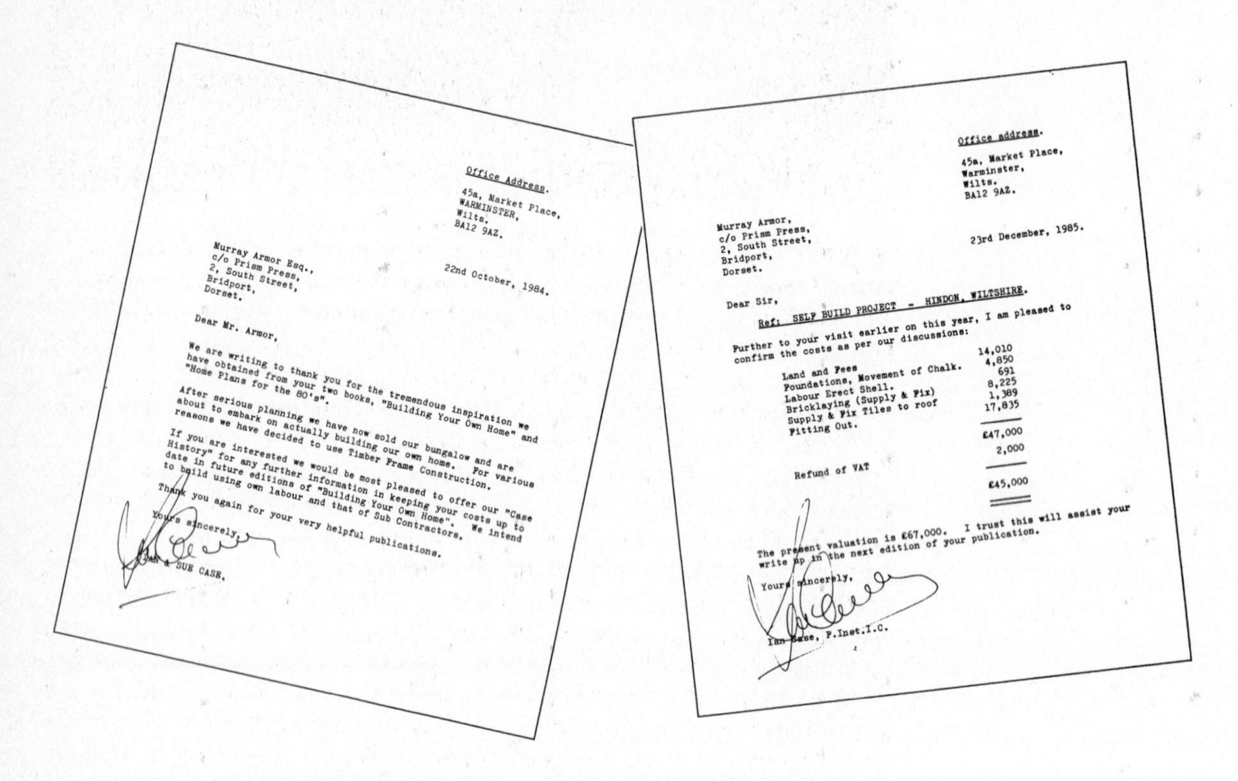

Office Address.

45a, Market Place,
WARMINSTER,
Wilts.
BA12 9AZ.

Murray Armor Esq.,
c/o Prism Press,
2, South Street,
Bridport,
Dorset.

22nd October, 1984.

Dear Mr. Armor,

We are writing to thank you for the tremendous inspiration we have obtained from your two books, "Building Your Own Home" and "Home Plans for the 80's".

After serious planning we have now sold our bungalow and are about to embark on actually building our own home. For various reasons we have decided to use Timber Frame Construction.

If you are interested we would be most pleased to offer our "Case History" for any further information in keeping your costs up to date in future editions of "Building Your Own Home". We intend to build using own labour and that of Sub Contractors.

Thank you again for your very helpful publications.

Yours sincerely,

IAN & SUE CASE.

Office address.

45a, Market Place,
Warminster,
Wilts.
BA12 9AZ.

Murray Armor,
c/o Prism Press,
2, South Street,
Bridport,
Dorset.

23rd December, 1985.

Dear Sir,

Ref: SELF BUILD PROJECT - HINDON, WILTSHIRE.

Further to your visit earlier on this year, I am pleased to confirm the costs as per our discussions:

Land and Fees	14,010
Foundations, Movement of Chalk.	4,850
Labour Erect Shell.	691
Bricklaying (Supply & Fix)	8,225
Supply & Fix Tiles to roof	1,389
Fitting Out.	17,835
	£47,000
Refund of VAT	2,000
	£45,000

The present valuation is £67,000. I trust this will assist your write up in the next edition of your publication.

Yours sincerely,

Ian Case, F.Inst.I.C.

Ian Case owns a mortgage and insurance consultancy in Warminster, and had spent much of 1984 negotiating to buy a very attractive site in one of the neighbouring villages. He eventually exchanged contracts in August and was fortunate in finding a buyer for his old house almost straight away. At the end of September the family moved into rented accommodation on a six month lease. Their plot had outline planning permission, but everything else had to happen in just six months. This was a tall order, but the Case family had two advantages. One was that Sue, who had once worked in the business, was able to return to work so that Ian could give the weekends and three weekdays to the job, and the other was that they chose Prestoplan for their timber frame and found that the companies local representative was someone whose enthusiasm matched their own and who was determined to make it all happen.

The first problem was getting the drawings approved in November as the local planning committee did not have a December meeting. Very wisely this was left to Sue, who could charm the birds off the trees, or in this case the planning officer away from his office and to the site where he was persuaded to

promise to hurry everything along. He was as good as his word, and all the paperwork was cleared in just fifteen days, which may be something of a record. Even so, bad weather and other commitments prevented a start being made before Christmas, and instead the time was used to make all the decisions on materials, fixtures and fittings, and to prepare a very long shopping list. Everything was then bought in one monster shopping spree at the Builders Merchants' sales in early January. This was something of an act of faith when the foundations had not even been opened, but a few days later a machine was able to get onto the site and started to dig a level plinth into the hillside. This involved excavating and carting away 600 tons of chalk before the foundation slab could be cast, and this was all done in just seven days before the timber frame was delivered on January 24th. With the help of Prestoplan this was up and the roof felted in just four days, and as the pre-glazed windows had been fixed in the panels before delivery the building was up, weatherproof and locked up in less than two weeks. By any standards this was quite an achievement, and it turned out to be essential to the timetable as the weather broke soon after. Indeed, such was the winter of 1984/5 that the house was completed inside before the bricklayers finally completed the skin to the frame outside!

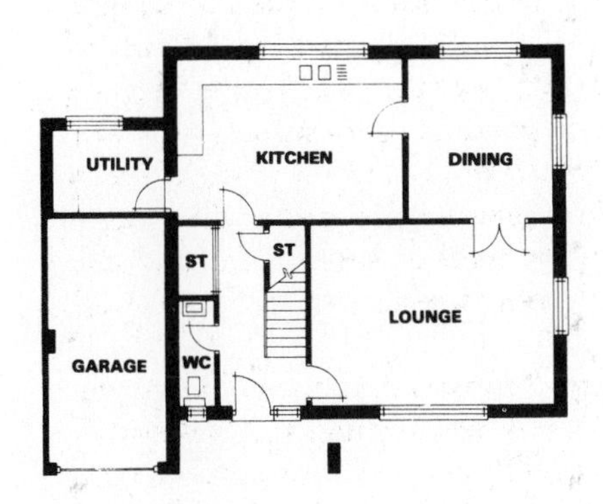

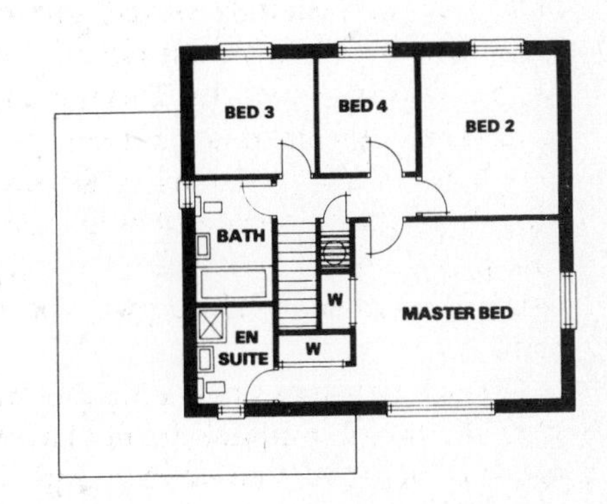

Tiling contractors were engaged to tile the roof on a 'supply and fix' basis, and it is interesting that the price for this was £100 less than the best price that could be obtained for the 'supply only' of the tiles. This is not unusual owing to the manufacturers' discount structure. Ian handled his own plumbing and electrical installation, and fixed some of the plaster boards. He had intended to do all of the boarding himself, using dry lining, but to keep to the timetable they brought in a plasterer and then adopted the plasterers' suggestion that he should skim all the walls at a contract price of £800.

There are four children in the Case family, and they all played their part in the cleaning out and decorating, and the family see the new home as something that they all built together. When they moved in on All Fools Day they had not only set what must be a self-build record but were also ahead of the statutory services; there was no electricity, and the water had to come from a temporary connection into a fire hydrant.

Advice for others? Ian had no doubt at all. "Plan ahead. Buy ahead. Never take no for an answer. And keep track of VAT receipts." The VAT bit is because in the hassle of the move the receipt for the bathroom fittings and the plumbing materials was lost, and because they had bought these for cash at a sale they could not get another and were unable to reclaim the VAT. A small problem in a terrific success story.

The total cost, including the land, was £45,000 against a valuation of £65,000 to £70,000. And looking ahead — "in two or three years we shall be looking around for an even better plot."

The Case family at the front door of their new home

A Multi-Level Home in Northampton

Keith Helliwell, a marketing executive who had to move to Northampton, and his wife Pat who has her own beauty business, were determined that if they were to move, it was going to be to the dream house that they had always promised themselves. Northampton has a development corporation to promote new industry and the new housing to go with it, and it leases homes on a short term basis to newcomers while they sort out where they are going to live. The Helliwell family — they had two teenage children at home — took advantage of this, and having sold their old home they looked very hard at the wide range of houses for sale. None were exactly what they wanted, but in their house-hunting they heard about serviced plots offered for sale by the Development Corporation, and sent for details. These included a prospectus for the sale of four plots on the crest of a hill, with a view that stretches for twenty miles. In the distance there was a lake, there was a spinney at one side, and the ground they were walking over was covered with red poppies. The sun was shining, and they decided there and then that this was where they wanted to live.

The plot that they bought was three quarters of an acre, and cost £22,750. Now all that they had to do was to get a home designed and built before inflation started eating into the money that they had received for their old home.

After a careful look at their options the Helliwells decided to approach D & M Ltd, who provide an integrated design and materials supply package for houses of any sort, anywhere in the country. They draw plans, obtain planning consents, and then supply the materials for the structural shell of the building all at one fixed inclusive price. The first stage was to settle on a design.

The slope at the top of the site was so steep that it was decided to look at the possibility of a multi-level design which would look like a bungalow at one end and a house at the other. The Helliwells knew exactly what accommodation they wanted, and that their budget would stretch to a building of 2000 square feet. The D & M architects had to consider the slope of the land, the need to arrange for drains and services, the planners requirements and the over-riding obligation to keep everything within the budget. There were a lot of meetings, a lot of sketches, and finally the plans were agreed and submitted to the Local Authority.

While the plans were being considered the Helliwells started to sort out all their options for materials, fixtures and fittings, and Keith began collecting dozens of quotations for every aspect of the work. Some sub-contractors, like bricklayers, quote for 'labour only' work, while others like plumbers and electricians quote on a 'labour and materials' basis. Besides getting quotes in these different ways it also made sense to go to see work that those concerned had recently completed, and to try to chat to the people for whom they had worked. As well as this there were prices to be sought for everything to go inside the house, as the D & M package is of the structural materials only. This suited the Helliwells very well as they had their own very firm ideas about bathrooms, the kitchen and much else. At this stage Pat said she found herself suffering from 'choice indigestion', with just too many options at too many prices, all of them having implications elsewhere in the grand design.

Pat Helliwell in the lounge, with the door to the family room behind her

By March, after six months of this, the plans were passed. The final design was for a home on three levels, with the entrance and living accommodation at the centre level, stairs up to a floor with the master suite, study and spare bedroom, and stairs down to a lower floor with the two children's bedrooms, another bathroom and the utility room. All this is shown on the plan, but what it does not show is that provision has been made for another large room in the roof to be completed at a later date, to be either a studio flat for one of the children, or a games room, or even an office as both Keith and Pat work from home. The stair well in the hall has been carefully designed so that the necessary stairs can be installed when they are required.

Work started on March 8th, and immediately there was a disaster. The excavator exposed an outcrop of rock which was quite unexpected. The experts quickly recognised that it indicated what is called a slip plane, which is a potential subsidence problem. D & M quickly designed special foundations to meet this situation, and the council gave their approval to these proposals with the minimum of delay, but the scale of the work required was daunting. Sixteen thousand bricks disappeared into the new foundation, together with seventy cubic metres of concrete, costing over £4000, none of which had been allowed for in the budget. It was a time not to panic, and the Helliwells didn't, reminding themselves that by managing the job themselves they were making sure that the problem was dealt with in the most cost-effective way. To console them, as the leaves appeared on the trees they found that the view from the plot was just as pretty as they had imagined it all winter.

After this difficult start everything went exactly according to plan. The care taken in choosing the sub-contractors to work for them was justified when the different tradesmen all turned up in sequence on time, and D & M kept the

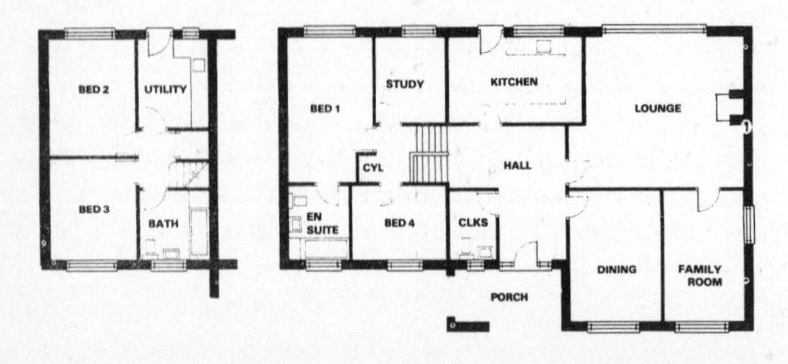

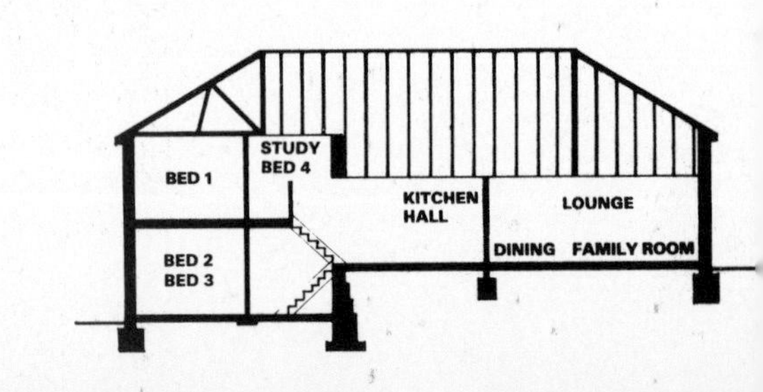

The Helliwells and their new home

deliveries of bricks, blocks, joinery, roofing materials and everything else flowing along. In fact everything went so well, and tempers were so good, that Pat decided to do what all the books warn against and changed the plans as the building was going up. The fireplace was moved from a position on an inside wall to an outside wall, and what had been planned as a very big dining room was cut down in size to allow for a small family room leading off from the lounge. All concerned took this in their stride, and the photograph shows the fireplace and the double doors to the family room, which is used for TV watching and playing chess.

Keith did most of the plumbing himself, making provision for a full air conditioning system to be installed if ever wanted in the future, although at present there is a conventional central heating system. He also laid the timber floors, as he was particularly concerned that there should be two inches of polystyrene insulation under the floor boards, and decided to handle this himself. Pat did all the decorating, and both of them worked endlessly at clearing up, stacking materials and making sure that everything was ready for the next day's operations. All this was in addition to their ordinary work, and, as they are quick to point out, building in this way is definitely not for the work shy.

And the outcome? They now have a dream home by any standards, and are busy turning their three quarters of an acre into a dream garden. They have planted 400 shrubs and hedge plants, made one patio, and are busy on another paved sitting out area at the bottom of the garden. A series of pools and a waterfall will come next year. Beyond that there is the studio in the attic . . .

The cost to date, including the double garage, has been £50,000 on top of the £22,750 spent on the land. The current value is a staggering £120,000. The high margin between cost and value is only part of the reward for all their efforts; the real feeling of success comes from having exactly the home that they wanted and from having made it all happen themselves, from the jacuzzi baths in both bathrooms, to the loop drive that avoids backing cars. The bad times — the foundation horrors, the time Pat fell though the ceiling and was a month recovering, the sheer slog of it all have become part of the total success. Above all, they found it all fun, and they called the house Poppies, after the poppies which were growing all over the site when they first saw it!

The Underwood Housing Association

The plots which the Sutton-in-Ashfield District Council made available to self builders in 1985 are described in an earlier chapter, and this case history is about a Housing Association which is building on twelve of them.

The Underwood Housing Association has only twelve members and is managed by one of the Wadsworth companies. It was the first self build group to be formed in the area, and both the managers and the members feel that they have had something of a struggle to get accepted. This is not unusual, but now that the success of this first group is assured, successor groups should benefit. Once an Association has proved that groups really do build on time, and within budgets, then the attitudes of local suppliers, solicitors, bank managers and others change quickly.

This Association got off to a slow start. A public meeting arranged in the summer of 1984 attracted only fifteen people instead of the usual couple of hundred, and only four of these seemed interested. However, the association was formed and eight other members were eventually found through newspaper advertisements and personal contacts. Exactly half of them had trade skills.

The original proposals for the scheme were for four bungalows, four three bedroom houses and four four bedroom houses, but it proved possible to change the mix so that everyone got the type of home that they wanted. The council agreed to sell the land for £80,000, and confirmed that work could start on site before the money was paid provided that interest was paid. Building costs were estimated at £29,100 for the large houses, £27,200 for the three bedroom units and £22,700 for the bungalows, with a projected saving of 25% on market value. These were June 1984 figures and the values were certified by a local firm of estate agents.

A start was made in March 1985, with a seventeen month programme drawn up by the consultants. Almost immediately there were problems.

The association had obtained a loan of a third of a million pounds from the Halifax Building Society. Once they received this the members started work on the site and wanted to buy the land, but legal requirements prevented them from actually handing over the money until they could get the title deeds. Unfortunately this was delayed by formalities concerning the road adoption, so that although the association wanted to complete the purchase they were unable to do so, and as they were working on the site they were clocking up interest on the purchase price at thirty pounds a day. It took over two months to clear the road adoption, and although the council waived £500 of the interest due as a gesture of goodwill there was still over £1,500 to pay.

The other problem was the weather and the nature of the subsoil. The spring of 1985 was very wet with rain almost every day. Although the site was well drained the soil was so slippery that it was impossible to walk on it without sliding about, and as these conditions continued into April the work fell behind programme and costs mounted.

Problems like these at an early stage will flatten some people, but simply provide a challenge to others. Mike Matthews, the Chairman of the association, decided that the answer was to cut the seventeen month programme to twelve months, thus saving interest on the loan from the Building Society and to spend this saving on employing sub-contractors. The

Mike Matthews, right, and two members with the first two bungalows completed

members agreed to this, and to increase the working hours with temporary electric lighting to enable them to work after dark. All through the summer they worked a minimum of 27 hours a week, with some members putting many extra hours of 'voluntary' work. This was all after they had put in a full days work at their various occupations. It paid off: when the photograph was taken at the end of September the first three houses were occupied, and the last unit will be roofed by November. The whole scheme should be completed in March 1986 in accordance with the revised programme, which is fortunate as no group of people could keep up the pace at which the Underwood association is working for more than a year.

Happily it looks as if the bad luck at the start of the project will be matched by good luck at the end, for local property values have started to rise steeply. It seems possible that the cost-value savings in March 1986 will be as much as 40%. It is being suggested that the rise in local house prices has been encouraged by the frenzy of owner-building at Underwood, of which the association is such a key part.

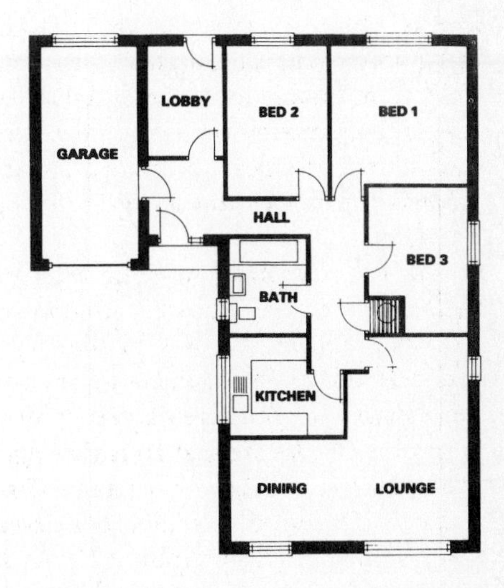

The Faulkener's House

There are two ways of buying a building plot. One is to look around and find one that is already for sale, and the other is to find the right piece of land and then to persuade the owner to sell it to you. The house in the photograph is in a very beautiful Somerset village where plots for sale are hard to find, and where Les Faulkener and his wife had to find the land and make the opportunity to build.

In their case it was through personal contacts that they arranged to buy the rear of an old walled garden behind a thatched cottage. It sloped steeply, and a new and expensive access had to be made, but it had superb views and the opportunity to buy it was the chance of a lifetime. When they first saw it there was an old wooden garage right in the middle, and what clinched the deal was an offer to build the owner a new garage to a high standard and to arrange for it to use the new access which they planned to build.

Designing a home for this particular site was a challenge, and the Faulkeners went to local architect Peter Saloman to prepare plans and obtain the planning consent. The final outcome is a most interesting home on four levels which is shown in the photograph. Behind the house you can just see the bank left after the excavation, and to the rear a patio window leads out to a walled back garden. To the front the master bedroom suite over the garage is reached from a landing half way up the stairs, and having the main bedroom on its own like this is a very convenient and attractive feature. Above all the house suits its surroundings and fits into the village as if it has been there for a very long time.

Once all the planning paperwork had been cleared the Faulkeners looked around for the best way to set about building, and decided to look closely at the package service offered by Design and Materials Ltd. After meeting their local Manager they made the long trip to Worksop to check out all that they had been told. Suitably impressed, they asked D & M to quote for the supply of structural shell materials for the plans which they had already got

Les Faulkener in the lounge of his new home

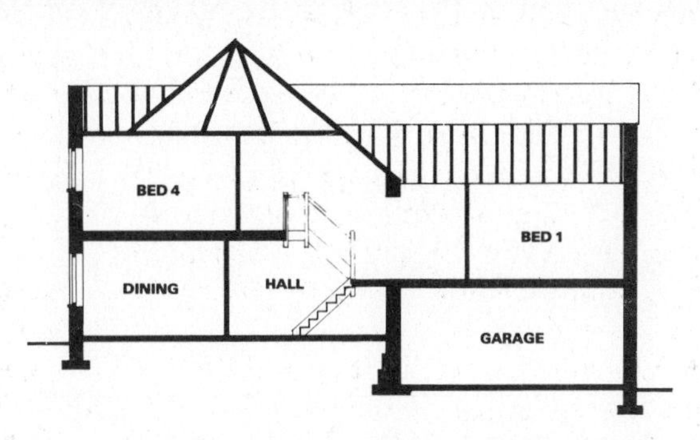

approved, and eventually went ahead on that basis. D & M were able to suggest some savings to be made by using their standard components, and this required a new Building Regulation which they handled without charge.

Work started on site with all the excavation and the elaborate groundworks handled by Les himself in his spare time, using hired plant. Sub-contractors were then used for most of the different trades. At the end of it all the total cost, including the land, was £62,450, against a valuation by a Building Society as soon as the house was finished of £100,000.

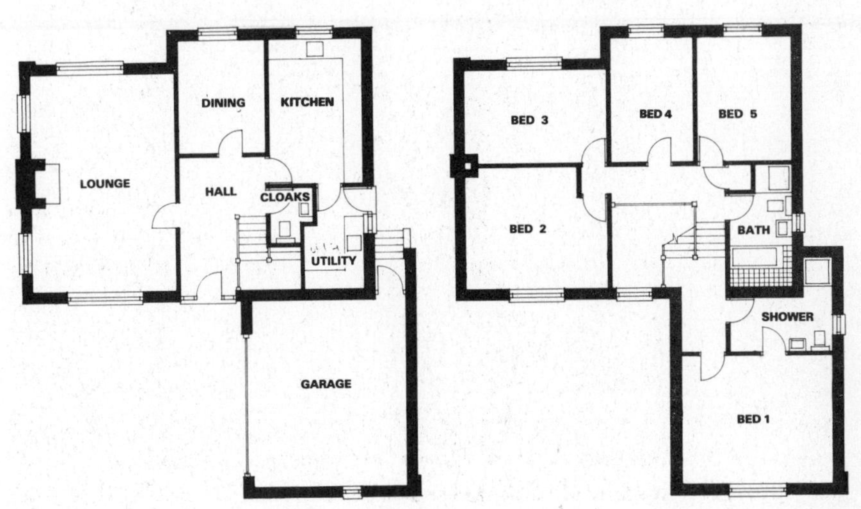

God Pulled the Strings

Most people are content to let their religious beliefs shape the decisions that they make in their lives: a very few literally follow a call, believing that the decision has already been made for them. The achievements of this latter group are often staggering, and, in this instance, involved a most impressive self-built home.

Doctor and Mrs Alston knew that they were called to move to East Sussex, and knew exactly where they had to go to establish a new medical practise. The Health Service has to give formal approval for new practises, and it took seven applications and four appeals to get the necessary clearances. By 1982 all this was settled, and the family moved into temporary accommodation in the area while they made arrangements for a new home. This had to be large enough for three children of their own, three adopted Korean orphans and two foster children, besides having a granny flat for parents. The budget was limited, but the family had no doubt that they would be led to know what was intended for them. There was total confidence that they would be led to a large field, with a wood and a stream, on which to build a new home. They even knew it would have a flock of sheep grazing on it when they found it.

A fortnight after they had realised this a patient in the surgery casually mentioned that he had a piece of land for sale. It was two acres, with a small wood, a stream — and when they went to see it there was a flock of sheep.

What it did not have was planning consent for a new house. An outline application was made straight away and was refused, so a further application was made for full planning consent for a home of 5,800 sq. ft, with eleven bedrooms, a granny flat and surgery. As with all else that they do, the family prayed long and earnestly for the success of the application. It went to the committee with a firm recommendation for refusal, and was unanimously approved without any reservations. God, the Alston family knew, had pulled the strings. All that remained was to build the house.

The drawings had been prepared by R.B.S. Ltd, from preliminary sketches drawn by Dr Alston, and the house was to be built with a timber frame. Work started on site in mid August 1983, and with the whole family working as a

Dr and Mrs Alston on the terrace of their new home looking out over the view that they knew that they were going to find

Some of the Alston family with their new home

team the concrete foundations were finished only four days after the digger moved onto the site. The only workmen employed were a bricklayer and a labourer. A friend who is a Minister took a few weeks holiday to help them nail the frame together, and many other friends helped on a temporary basis from time to time. R.B.S. tiled the roof (which is part of their standard contract), and patients and new neighbours helped with advice and the loan of equipment. A retired builder helped with the tricky aspects of the drain connection, and Doctor Alston himself put in sixteen hours work a day for eight months until the first rooms were habitable in April 1984. The last of the family moved into the house from four caravans on the site in August, although work to finish everything off went on until 1985.

Building costs were £18 sq. ft. which in 1983 were the average self-builders costs advertised in that year's edition of Building Your Own Home. This included £55,000 for the R.B.S. materials and service. It also covered the expensive site works involved in building on a sloping site, which has enabled a six car garage to be built below the ground floor, which is of beam and block construction — all the beams lifted into position in one day by the whole family working as a team.

Everything about this job is larger than life; the house itself, the whole concept, the whole story. But as the Alstons insist, they were taking their orders from a very special foreman, and with His help everything was bound to turn out right.

A Countryman Home in Gloucestershire

Farmers are great self-builders at both ends of the scale. Those who are setting up on a new holding often build for themselves as the best way of getting a farmhouse on a tight budget, while those who are well established often decide that their existing arrangements for the maintenance and other routine building work on a successful farm can easily cope with building a new house or bungalow. The cottage featured is in this latter category, and is being built by a landowner whose concern is that it shall be exactly right, and who is managing the job himself to make sure of this.

The farm is in a particularly attractive part of Gloucestershire, and outline planning consent was only obtained because a demolition order was served on a very old cottage which was quite beyond repair, so that the replacement dwelling was approved on a 'one for one' basis. The farmhouse and all the other buildings in the area fit into the landscape as if they grew there, and the primary requirement for the new cottage was that it should suit the site perfectly, with no compromises at all. This is a difficult thing to arrange, as people often have different views of what is appropriate to a sensitive site, but in this case a visit to the Homeworld exhibition at Milton Keynes provided an immediate answer. One of the show houses at Homeworld was a timberframe cottage by Countryman Homes of Oxford and it was exactly what was wanted: indeed, on the road to the site in Gloucestershire is a cottage of exactly the same external appearance that has stood there for 300 years!

Countryman Homes offer a typical timberframe service which covers design, handling planning and building regulation paperwork, the supply of the timber frame itself, and help with finding contractors. In this case an interesting feature of the job was the existence of the cellar of the demolished cottage. After considerable discussion it was decided to retain this under the new building, even though the cellar walls had to be rebuilt to meet the current Building Regulations although the old ones were completely sound and waterproof. The cost of this was an extra £3,000 but the new cellar is considered well worth it.

Two outside contractors were brought in at the suggestion of Countryman Homes, one for the foundation work and one to erect the timber frame. Everything else has been done by local tradesmen, and the general opinion is that the local people could have easily done the specialist work which was done by the outsiders as the Countryman drawings were particularly well detailed and easy to follow.

An unusual feature of this project is that at the time that the photograph was taken in September 1985 it had not been decided who is to live in the cottage, and it was really being built to 'use' the planning consent before it expired. As a result there has been no hurry over the job, which has gone on for eighteen months on an on and off basis whenever tradesmen are available from more urgent work. One or other of the farm staff will move in when everything is finished. This is a very different situation from most self-build operations.

The materials were selected with great care. The walling is in handmade bricks from a demolition, and they were probably Victorian. However, some of the 2″ Tudor bricks from the old cottage were found to be sound enough to be reused, and were laid as a stringer course around the cottage at first floor

The house under construction, showing the band of narrow tudor bricks and the dentil brickwork at the eaves

level. This is one feature of local buildings, and others are the brick arches over the windows and the dentil brickwork at the eaves. The roof is clad with second hand Rosemary tiles. Even the doors are from a demolished building dated about 1830, so that they are in pitch pine and will last for ever.

The result is a charming home, with a cellar and a well at the back door, which combines the character given to it by the old materials with modern amenities and a high standard of insulation. The total area is 1,448 sq. ft., including the cellar, and the final cost will be within £40,000. This gives £27.62 sq. ft. in a rural area where £35 sq. ft. is considered the usual cost of building a new home.

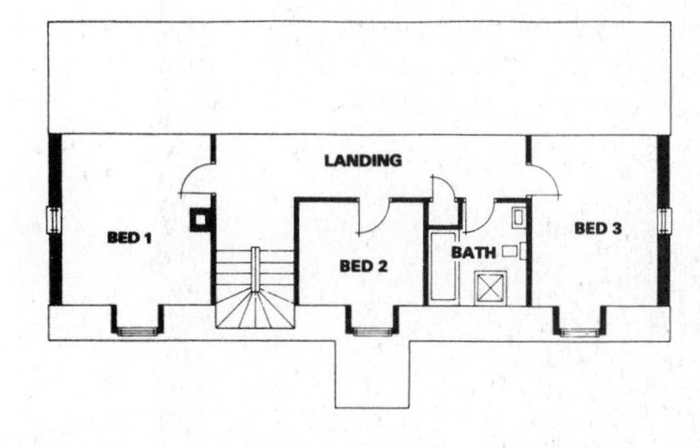

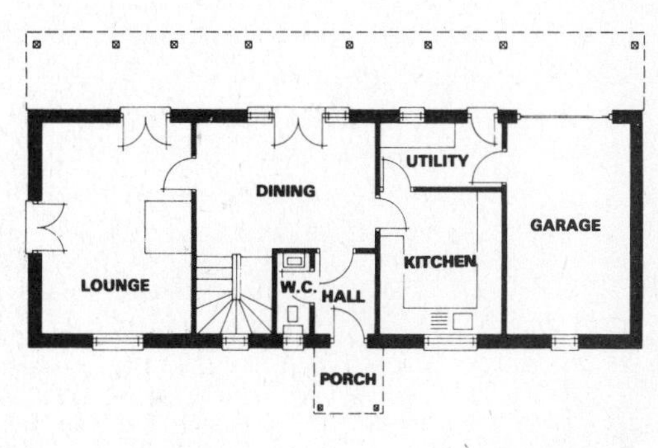

Sylvan Hill— An Urban Self-Build Association

This is an important self build scheme for a number of reasons. First of all it is in a London Borough, building in an area where the demand for self-build housing is very high and opportunities for it are very rare. Secondly, it is successfully dealing with very dificult ground conditions which had previously deterred big development companies who had looked at the site. Finally it is achieving a very satisfactory cost to value ratio. It all demonstrates that self-build really has got a role in the cities, and Sylvan Hill may encourage provision to be made for those who want to build for themselves in urban renewal programmes.

The site is directly between the two Crystal Palace television masts, and was a heavily wooded piece of land on a 1 in 5 slope. It was council owned, and although zoned for housing it had escaped development as it needed elaborate site works to deal with the slope and to protect the mature trees.

This was a typical 'managed' project, with consultants Wadsworth and Cudd arranging to buy the land, finding finance, and making all the arrangements for work to start before any steps were taken to form an Association. In August 1983, when all this had been done, they put a notice about the project in the *Croydon Advertiser*, asking for prospective members to send for an application form. In this instance they did not hold a public meeting, and interviewed applicants privately. The first time the members met each other was at the inaugural meeting of the Association. All details had been arranged, including finance from Barclays Bank, and a start was made on site at Christmas.

There are 22 units in three blocks, with 2 different house types. All of them are traditionally built 3 bedroom, 3 storey town houses, and they have been carefully designed by the architects to suit the site. Two level areas have been carved out of the slope, and two of the blocks are at one level with their own entrance, while the third is at a higher level with a separate entrance which is retained by a crib lock retaining wall over 20 ft. high. This is a sloping wall with slots through to the ground which will be planted with shrubs, and the

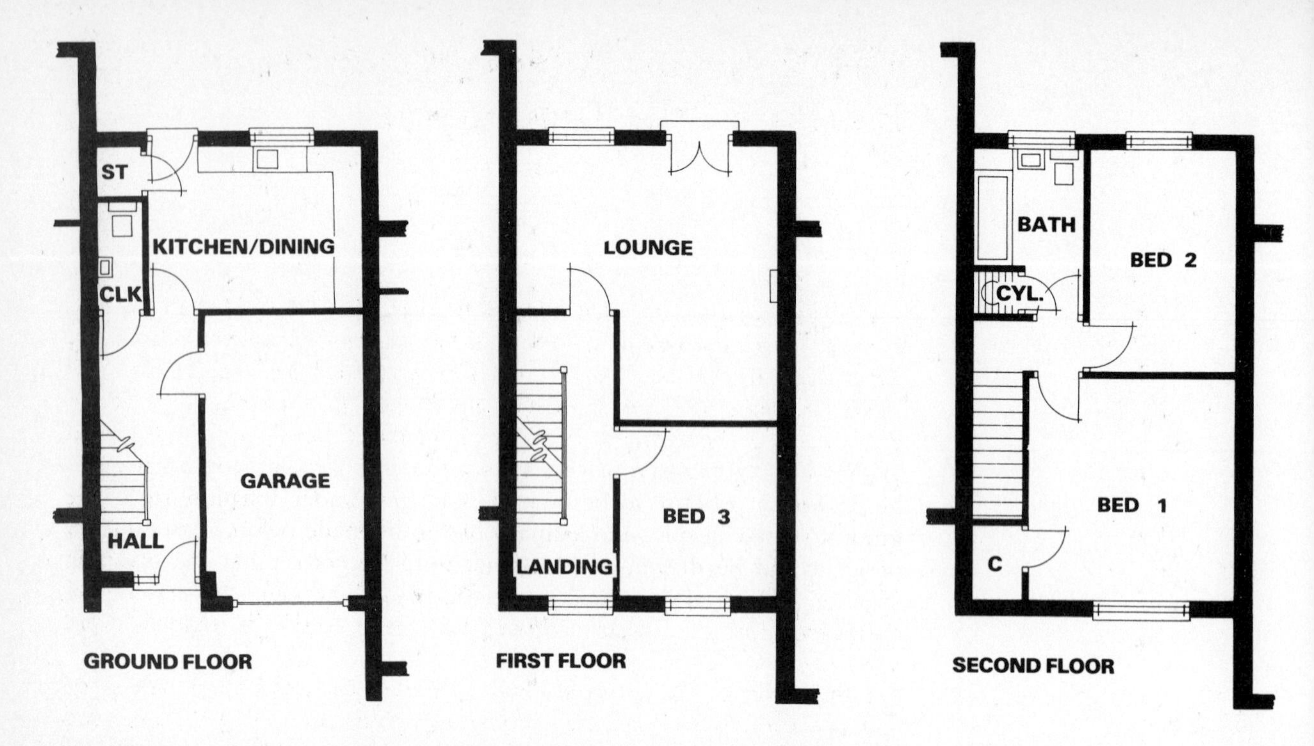

landscaping of this wall is to be a key feature in the appearance of the completed site. Great care has been taken to protect the trees, so that the heavily wooded character of the area will be retained, but in spite of the best intentions there have been disputes about this with local conservation societies and others. Fortunately all this has now been resolved.

The land cost £180,000 and the target cost per unit is under £30,000 when finished, against a projected 1985 valuation of £55,000. All the work in the foundations is out to contract, as is the plastering, but everything else is being handled by members.

The first houses were occupied in September, 1984, and there are hopes that rising property values will lift current valuations significantly by the time everything is finished. The cost breakdown is given below. This is believed to be the first self-build scheme in the Croydon area. Let us hope it leads to others.

Costings

	4 Type BE	*7 Type B*	*4 Type CE*	*7 Type C*
Plot cost, Legal costs, Site works, Services,	12379.44	12712.35	12441.35	12754.43
Plant purchase, Services connection charges	496.36	496.36	496.36	496.36
Foundations	1360.73	1285.52	1381.89	1306.68
Superstructure	2933.99	2013.29	2847.60	1947.30
Roofing	802.23	802.23	802.23	802.23
First and second Fix	3149.33	3149.33	3152.08	3152.08
Final Fix	1539.92	1539.92	1539.92	1539.92
Loan interest, and consultants' fees	5874.00	5874.00	5874.00	5874.00
Total Development Cost	£28536.00	£27873.00	£28536.00	£27873.00
Market value (1983)	£43000.00	£42000.00	£43000.00	£42000.00
Cost saving	£14464.00	£14127.00	£14464.00	£14127.00
Percentage saving	33.637%	33.637%	33.637%	33.637%

The Pattisons: A New Home at the Bottom of the Garden

Until last year Keith and Liz Pattison, both of them teachers, lived in a period cottage with a large garden in a village just to the east of Bristol. When a newly developed cul-de-sac provided a road frontage to the bottom of the garden it offered an obvious opportunity to build a new house for a family that was growing out of the cottage anyway. To see what could be done they contacted Design and Materials Ltd in July 1984, and made arrangements for plans to be drawn for a house to be built in Bradstone under a pantile roof. The ground sloped steeply, and a survey had to be made before a preliminary design could be drawn, so it was not until December that the planning application was made. Drainage arrangements and other aspects of the application had to be negotiated, and then a start on site was made in April 1985.

Problems arose at an early stage. Hundreds of tons of soil had to be excavated and moved off the site to give a level plinth on which to build, and right at the end of this operation they discovered a seam of coal which ran under a corner of the house. This had to be excavated to a depth of twelve feet and the hole had to be backfilled with concrete, giving a thousand pounds of unforseen expenditure. Then, when all this was dealt with, the foundation walls built up, and the foundation filled ready for the floor slab the Council stopped the work while consideration was given to the floor levels. Their concern was the relationship of the height of the roof of the new house to the roofs of neighbouring buildings, and there were various different aspects of this. The Building Inspector had been very involved with the coal seam and had verbally approved all that had been done, but following local concern about the appearance of the new house, and a suggestion that it was going to seem to overshadow its neighbours, the council's senior staff became involved. The outcome was that the foundation walls had to be lowered by

Keith and Liz Pattison, with a friend helping with the glazing in the background

Tiling nearly completed — the stage when it can all be seen to be coming true

eighteen inches and much of the expensive fill so laboriously shovelled into the foundations and consolidated had to be dug out and taken to the tip. The extra cost was another £1000!

After this sort of start a little despondency would have been expected, but the Pattisons took it all in their stride and pressed on. The brickwork was let to a local bricklaying gang at a lump sum price, which included helping Keith fix his own roof ready for D & M to come to fix the felt, battens and tiles in accordance with their standard contract arrangements. Next a father and son team of a plumber and an electrician started work on their sub-contract. As is so often the case with self-build, the tradesmen were found through personal contacts. One was the husband of the leader of the playgroup attended by a small Pattison — and this detail is not irrelevant because it demonstrates the way in which so many self-builders use their contacts to the full. Talk at the playgroup about the Pattison house led to another mother offering her husband's skills as an artex specialist, and so it went on. Invariably contacts of this sort are more useful than the yellow pages.

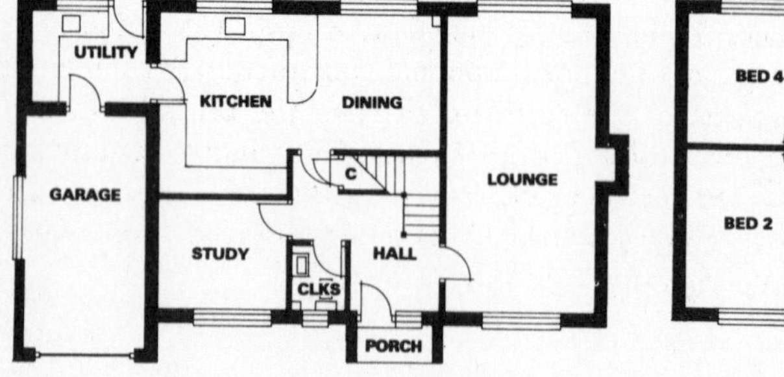

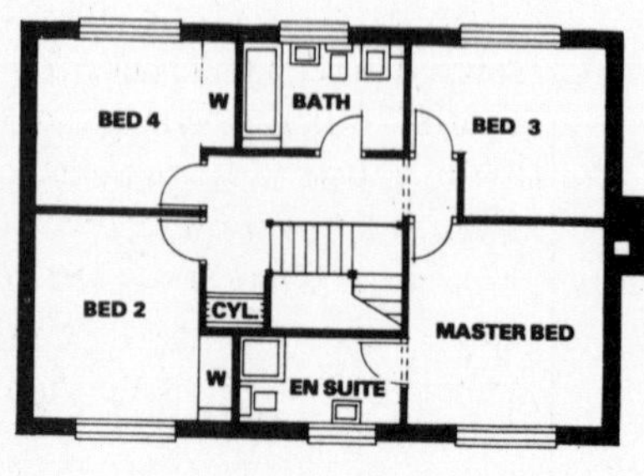

Cost Breakdown	
D & M for plans	£199
Architects Certificates	£350
Insurances	£170
Site clearance	£250
Readimix concrete	£1,111
Blocks for footings	£375
Extra foundation costs	£1,005
Oversite concrete and steel mesh	£388
Plant hire	£182
Bricklayers	£3,850
D & M package, first stage	£12,951
D & M package, second stage	£8,342
Electricity connection	£150
Gas connection	£220
Water connection	£200
Drainage	£369
Sand	£210
Scaffold hire	£316
Electrics, materials	£580
Electrics, labour	£250
Floorboards	£200
Plasterboard	£300
Plumber & heating, labour and materials	£2,910
Plasterer	£1,100
Carpenter	£1,040
Doors, wardrobe fronts	£872
Kitchen	£1,400
Artex	£200
Paint	£109
Other costs	£2,692
Total	£42,291

Two different retired builders who lived locally were persuaded without any difficulty to change their role from spectators to being expert advisors. Before long they were actually working on the job, and built the fireplace, installed the patio window (always a tricky job), and constructed the garden walls. One of them, at over 70, proved to be the only person who could excavate the drainage trenches in the local rock without losing his temper. Howard and Percy are their names and this case history was authorised by the Pattisons on condition that they got mentioned in the text. Good for you, Howard and Percy! The house was occupied in August 1985 after four months, which is very good going for traditional construction even without all the foundation problems. The costs are given in the breakdown, and include all the appliances for the super new kitchen. The value of the plot if it had been sold on the open market would have been about £18,000, and the value of the new house is between £70,000 and £73,000 in an area where property values are moving up very quickly.

A House in a Yorkshire Village

This is another story of a couple who built a home after reading an earlier edition of *Building Your Own Home*. The case history of the first new home that this book had inspired appeared in the third edition; when writing this seventh edition it would have been easy to fill the whole book with narratives like it! At any rate, this story starts with Terry and Gill Baker buying a third of an acre site for £12,500 in a village just east of York, and writing to Design & Materials Limited because they read of them in *Building Your Own Home*. After meeting one of the Company's field staff at the site they were sent an estimate for architectural work, which is always the first stage of any contract to engage a company offering a package service. The wording of the estimate was important, as there was no firm obligation to use the material supply service when planning consent had been obtained, but full architectural fees would be involved in building the house without using this service. All of this was explained very carefully at the time.

The site was set in a wood on the outskirts of the village, and although there was an outline planning consent it was anticipated that there would have to be detailed consultations with the planning authority. The application for approval of reserved matters was made on February 17th, and on March 19th a letter was received from the authority asking for a meeting on the site, and suggesting that the applicants brought along samples of the stone and tiles which had been proposed in the application, and also other types of stone and tile which the planning officer wished to discuss with them. The meeting was a success—planning officers are much more approachable when away from their desks—and the materials agreed were Bradstone traditional walling in the weathered cotswold colour, and Marley Mendip Mossborough Red roof tiles. The joinery was all to be stained, which suited the Bakers who wanted to use mahogany doors and window frames anyway.

Meanwhile a tentative budget had been put together which totalled £26,762, but which did not include any allowance for the cost of getting electricity or water to the site. This all seemed practicable, and with Barclays Bank offering both building finance and a mortgage when the house was built, a start was made on Good Friday. A lot of family help was available to clear the site and to bring in a great deal of soil which was to be used to build

Terry and Gill Baker at work.

up the ground level around the home when all was finished. This involved an additional height of foundation walling below floor level, but was to make a lot of difference to the appearance of the finished home. Thereafter the Bakers did everything themselves except for the bricklaying and the plastering, which was done by sub-contractors. They poured the foundations, cast the slab, laid the drains, carried materials for the bricklayers, fixed the roof, did all the joinery work, the plumbing, the electrical work, the glazing and the decorating. The worst job was the drains, working in very heavy clay in an exceptionally cold November and December. Design & Materials had fixed the roof tiles as part of their service, but the amount of work that the Bakers did was still prodigious for a couple who were also working at their normal jobs, and had only the evenings and weekends free to work on the new home.

Everything was finished just in time to move in and eat Christmas dinner in the new house, among half unpacked packing cases! The total construction time had been eight and a half months, and the total cost was £27,560, which is £900 more than the original estimate. One reason for the budget over-run was the high cost of providing services to the site, and the expense of this caused more worry than anything else. The electricity board had quoted £500 for their supply line, and then sent a bill for £1040. The water board quoted £90 for a connection and then submitted an account for £200. A month later, and after lengthy arguments, the bills were reduced to the figures shown in the cost breakdown opposite.

One of the conditions of the help given by Barclays Bank—and the Bakers described their Bank Manager's co-operation as "fantastic"—was that they had to obtain architect's progress certificates as the work progressed. This could have been a problem, but their solicitor put them in touch with a lady architect who passed the site twice a day while travelling to and from her office, and who undertook the inspections at a very modest fee. The only other problem that had to be dealt with was that the cavity trays were forgotten. These were special damp courses running across the cavity at the point where the low level roof joined the gable wall. As the bricklayer had forgotten about them the wall had to be opened up and the trays painstakingly inserted section by section. Nothing else gave any trouble.

Talking to the Bakers one gets the impression that the way in which the whole job went so smoothly probably owes a lot to their easy manner and obvious knack of getting the best out of anyone working with them. In some ways this is the most valuable attribute a self-builder can have.

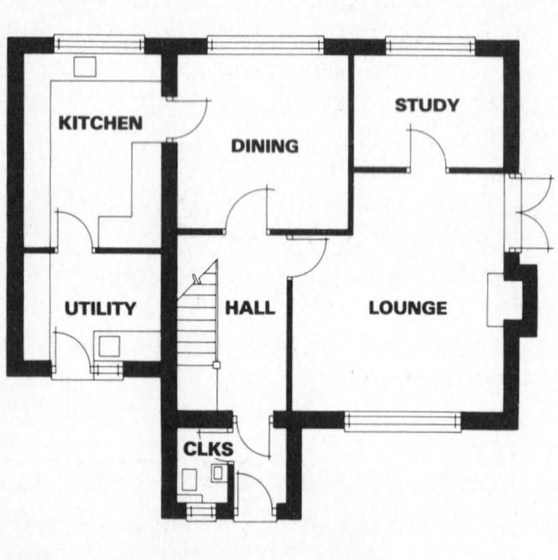

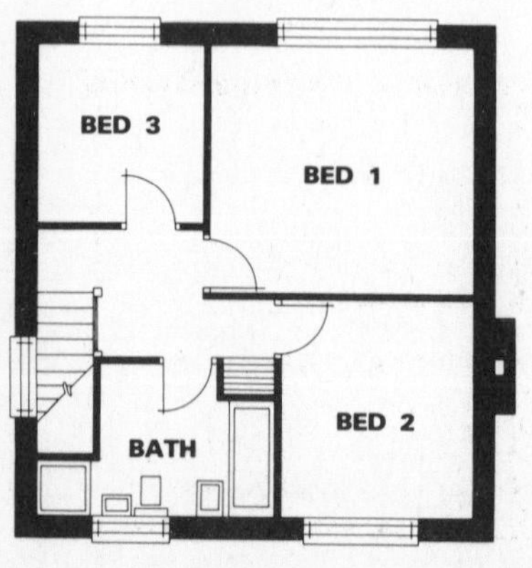

Planning and Building Regulation fees	104
Building Regulation Inspection fees	128
Services: Water	101
Electricity	665
Footings — ready mix	310
Hire of Mixer	50
Scaffold hire	245
Stone fill	350
Bricklayer from dpc (labour only)	2400
Plasterer inc. cove to lounge (supply & fix)	1650
Plumber inc. leadwork (labour only)	625
Sanitary ware for bathroom & cloakroom; radiators, room heater — Parkray 111 GL	1200
Electrical — wiring, light fittings etc.	575
Kitchen & Utility units	590
Oven & Hob	520
Bedroom cupboards & drawers	270
Tiles — kitchen & bathroom	150
Woodstrip floor to dining room	165
Paint & Varnish	70
Septic tank and drains	780
Site Insurance	100
Architect's fee for visits etc.	125
Builders Merchants — inc. foundation bricks, sand, gravel, cement, copper pipe, plumbing fittings, lead, floor insulation	2655
Timber Merchant — inc. ground flooring, worktops and kitchen sink, timber ceiling to hall	1185
D & M Package	15363
	30376
Less VAT refund	2816
	£27560

A Timber Frame House at Carnforth

In May 1983 the author received a letter out of the blue from Mr Peter Lord, who wrote that he was building a timber frame house to a novel design, and that it might make a good case history for *Building Your Own Home*. It has, and this is the story.

Peter and Alison Lord live in Carnforth in Lancashire, where he is an Environmental Health Officer with the local authority. Self build is popular in the area and Peter had friends and colleagues at work who had built for themselves, both on their own and as members of self build groups. The local council sells plots to private individuals by tender, and at one time the Lords had tried to secure one of these themselves, without any luck. In 1982 they decided that it was really time to get involved with all of this, and as a start they visited the Homeworld exhibition at Milton Keynes, and also went to East Grinstead in Sussex to see a development of timber framed Scandinavian homes under construction. This visit convinced them that a timber frame home of this sort was what they wanted.

Soon after this a plot of land was advertised at the holiday resort of Bolton le Sands, between Carnforth and Morecombe. It was near to the sea, sloped steeply, was heavily wooded, and had existing planning consent for a dormer bungalow. The advertised price was £12,000 and the Lords offered £10,400. This was refused out of hand, and then a week later the estate agent got in touch again to accept. They had got a site.

The next stage was to submit drawings for the Scandinavian design which they particularly liked. It was so exactly what they had always wanted that they had never considered that it might not be acceptable to the planners, but this proved to be the case. They were advised to re-apply with a design that "was better suited to the site" — or, of course, they could have built the dormer bungalow for which there was existing consent.

At this time a friend introduced the owner of a small firm that specialised in erecting the frames of timber framed houses, and he in turn introduced Lancaster Saw Mills who would design and supply a kit for a home to suit both the site and the planners. This they did, and everything was sorted out for a start in September 1982. Things were then postponed until January 1983 as the sale of the Lord's existing house was delayed.

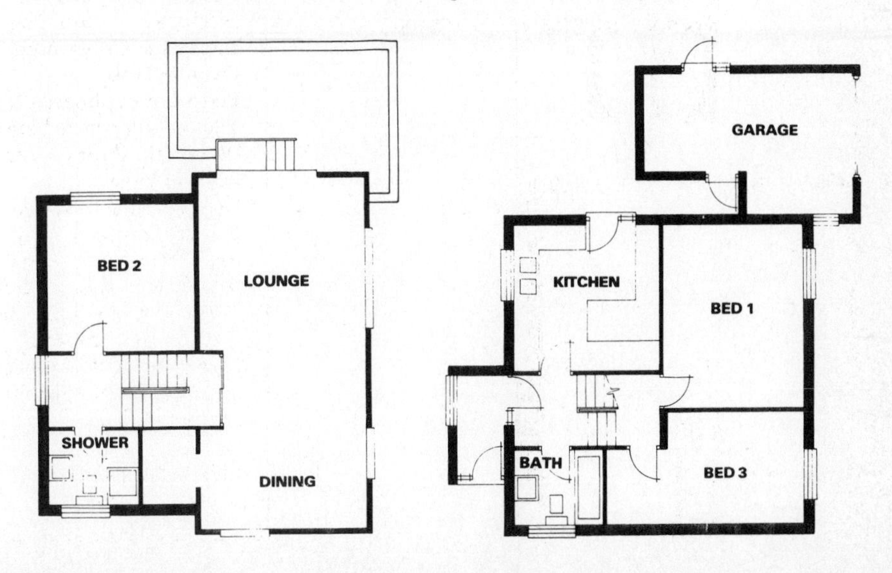

Peter and Alison Lord's house at Carnforth.

Meanwhile Peter Lord had set his budget for the building work at £30,000, and started to sort out the finances. The building society where he was an investor was less than encouraging, so he changed to the Anglia where he was made to feel very welcome. A contingency overdraft was arranged with the bank, and a building surveyor engaged to provide Architects' Progress Certificates. All these arangements worked well, and the Anglia Building Society in particular was always most helpful.

A start was finally made in January 1983, when the trees were cleared and a surveyor friend put in the four corner pegs. These provided a datum for Peter to set out the building, and the foundation trenches were dug with a hired JCB. Some of them were very deep as the ground sloped steeply.

The footings were to be poured at 9 a.m. on a Saturday morning. On Friday it rained all day and all night, and dawn on Saturday saw the Lords bailing out the trenches with buckets: by the time the mixer lorries came the trenches were more or less dry; they were exhausted, and all the concrete had to be laid. With the help of the digger driver somehow it all got done.

A gang of local bricklayers were found for the next stage, at a price of £90 per thousand facing bricks laid, £50 per thousand for commons, with an hourly rate for other work, and the complex foundations were quickly completed. The timber frame erector then took over and put up the shell for £1700 which included fixing the windows, doors, and all the cladding. The frame itself together with other materials from Lancaster Saw Mills cost £13,000. At the start of March it was all up, and although very heavy and persistent rain delayed tiling the roof it was possible to get it lathed and felted so that work could continue below. By Easter the brickwork was completed, and the shell was thoroughly weatherproof.

To fit out the building the Lords engaged an electrician who lived across the road, a plumber who worked for them at weekends, and a carpenter, managing to get all the work done at very reasonable rates. As is usual with timber frame construction the house was to be dry lined, with the plasterboards jointed using special techniques instead of the traditional wet plastering. Peter found that British Gypsum ran three day courses about this at Carlisle, and managed to attend one of them. This was a great success: he was given the kit of tools he needed for the job, and to get him started one of the experts from the firm's technical services department came along and worked with him for a day or so. This is typical of the help which the building industry gives to self builders who approach situations in the right way, and it is always nice to hear of it.

The interesting design is more easily understood by reference to the plans than from any description, and when working out how the accommodation is arranged, remember that the sides and back of the house are set in thick trees, with a super view to the front. There are boarded ceilings to the lounge, dining room, kitchen and both bathrooms, and this matches the style and appearance of the Rippers Alpha windows. The high standard of insulation includes double glazing throughout.

The area is 1,285 sq. ft., and the building cost was £35,495. This has to be considered in relation to the complex design and the difficult site. The total cost, land plus building, was £45,895, and the valuation when it was completed in December 1983 was £57,500.

Peter Lord exercising his dry-lining skills.

COSTS:

Land	10400
Foundations, removal of soil, brickwork to DPC	2500
Timber frame kit	13600
Joinery	3400
Brickwork — house & garage (including correction of substandard work)	3800
Roof & garage roof	1300
Scaffolding hire	550
Fireplace, chimney & wood stove	1050
Plumbing, central heating, bathrooms & tiles	2800
Electrics	550
Kitchen	2200
Drainage	450
Plastering to chimney breast etc.	140
Decoration, wood preservatives etc.	460
Rainwater goods & fitting	175
Drive, garage floor, paths etc.	400
Wardrobes	450
Theft	200
Services	320
Fees	600
Hire of equipment	50
Miscellaneous	500
	45895
Approximate value of house	57500
Cost per sq. ft. £25 (approximate)	

N.B. All prices are rounded up/down to nearest £5. They do not include VAT.

Su's Diary

In August 1984 the author visited Clive and Su Mortley, who he knew were just finishing a new home which they had built themselves with a great deal of help from their families and friends. He settled down in their caravan with notebook and pencil to ask questions, and then found that whenever Su was uncertain of the sequence of events or of a date she kept referring to a diary. One look at it, and the interview quickly changed into a successful plea to permit the diary to be published.

It is the story of a house in Kent, just as it was written up alternately by Clive and Su at odd times as the job was going on. It fills two large manuscript books and is illustrated with dozens of colour snaps—which will not reproduce on these black and white pages. It carries the real feel of the job—despair, determination and triumph. Here it is—two self-builder's private thoughts made public.

At the beginning of November 1982 we (or rather Clive) decided in principle to build our own house. The idea was chewed over for about a week, but refused to go away, so the only course open was to turn the idea into action. Our time-table for those early months was as follows:

13th November: saw building society to outline our proposals. They seemed happy and told us to come back when we wanted to apply for a building mortgage.
15th November: Estate Agents came round to measure up our present house.
18th November: House went on the market.
19th November: House sold—deposit put at estate agents anyway!
26th November: saw bank manager to outline our requirement for short term bridging finance.
30th November: Found the new site and paid deposit.
1st December: Instructed building society, bank and solicitor to proceed with purchase of land and arranging of finance.
2nd December: Posted card to Design and Materials to arrange for someone to come and see us.
7th December: David Snell from D & M talked us through our ideas and looked at the plot of land. As a result of this meeting, plans for the 'Bakewell' design were to be drawn up and submitted to the local authority for detailed planning permission.
15th December: Received plans from D & M. Small query re size of land at rear of plot as ordnance survey seemed to indicate that it narrowed. Submission to authorities delayed whilst this was clarified.
12th January: Signed contracts on our present house and set completion date of 28th January 1983. Contracts exchanged the following day.

So, on 28th January we moved out of our home to Clive's parents's house. We had asked them if they wouldn't mind a couple of lodgers (3 including the cat) for a period of no longer than three months. We were unable to move into anything onto the land as it was not yet ours and we had nothing to move into at that stage. The application for detailed planning permission and building regulations were also made on the 28th.

On 19th February we went to look at a caravan Clive had heard of going cheap, and offered £200 for it. On 22 February we were the proud owners of a very dirty and smelly caravan. As it was going to be our home for many months

The diary itself, with the colour photgraphs which unfortunately will not print in black and white.

while we were building the house we decided that we would spend some time and money on the inside and it wasn't ready for living in for some time. There was no great rush to finish the caravan because we did not actually own the land until 21st March and permission to put the caravan there did not come through until 13th June. In addition we were unable to lay the drainage to the caravan (which would later act for the house) until we had detailed planning permission. This came through at the same time as that for the caravan. The three months we had originally anticipated had now become five months and we moved into the caravan on 2nd July.

The timetable for the supply of all services was as follows: Water laid 2nd June—Electricity laid 20th June—Telephone installed 27th August—All were requested on 29th March!

Whilst we were waiting to put the caravan on our land we stored it by some farm buildings next door. This enabled us to have an electricity and water supply so that we could carry out work on the inside. We moved the caravan onto our land on 14th April.

Marking out for the house started on 20th June with assistance from David Snell of D & M, and after a short holiday break to strengthen us, digging finally commenced on 30th June. We had the two lorries and one JCB and the whole job including the hole for the septic tank was finished in two days.

Next the foundation trenches had to be tidied and levelled, with datum pegs to show the height of the concrete for the foundations. At the same time the septic tank and drainage had to be installed as this was desperately needed for the caravan. We now begun to realise just how busy we were going to be in the ensuing weeks until the bricklayer came in and we could take a back-seat for a while. The situation on the septic tank was serious—without it the toilet in the caravan could not work and the arrival of the strawberry picking season in the fields immediately behind us meant that it was no longer that easy to 'nip behind the caravan'. However, despite Su's protests, the levelling of the trenches was completed first because Clive argued that once the concrete was in the bricklayer could start and, he could install the septic tank whilst the foundation bricks were being laid. So, on 10th July the trenches were finished and the septic tank was lifted into its hole. The caravan was plumbed into the tank on 21st July—19 days after moving in. If it had not been for a kindly soul lending a chemical toilet life would have been unbearable. It was unfortunate that the day was very hot and very sunny which meant that the concrete was going off before it could be moved round the trenches. Also the heat gave the problem of the concrete cracking whilst it was drying so it had to be kept wet and cool.

Tony, our bricklayer, started work on the foundations walls on 15th July and finished them to dampcourse on the 22nd. The same day our first delivery from D & M arrived and that weekend was spent moving the materials to the right places on site. The two loads of facing bricks had arrived earlier in the week and little did we know at that time how much of a problem they were going to be. They appeared OK to us and we did not bother to open any of the packs until they were needed — and that was the first mistake we made in our attempts to build our own home.

Our drain run went through the floor of our garage and the next stage for us was to get the drains put in. Clive decided to do this himself and it certainly tried his patience. Drains have to fall at a 1:40 gradient and normally you would simply start from the house until you got to the position of your septic tank and then lay the tank to that depth. Because we had laid the tank first we had to try and juggle the required fall between the house and the tank. We had worked it out first but in our calculations we had assumed the land to be level because it looked level. That was the second mistake. The land dropped away towards the tank and on 30th July we admitted to ourselves that we were 9½ inches short on the required drop. This posed the biggest headache to date and at one stage we even considered digging up the septic tank and resinking it to the correct level. It was only the thought that we might damage the tank that stopped us doing this and we eventually decided to bring the drains up at the house end. This gave us slight concern as to whether we would have enough clearance over the drain pipe for the necessary amount of hardcore and concrete for the garage floor, but it worked out with millimetres to spare.

The bulk of the drainage work was finished on 12th July and Clive had already turned his attention to filling the foundations with hardcore for the oversite. After a big hint from the Building Inspector we decided to use reject ballast and 60 tons disappeared into the foundations. That took some shovelling and barrowing. Concreting started on 17th August and this time it was done by hand after the lesson learnt with the footings. Also it meant that we could work in the evenings.

Once the hardcore is in it has to be compacted using a vibrating plate, and this was hired. It took two attempts to get this stage of the proceedings passed by the Building Inspector as the gap from the hardcore to the top of the brickwork has to be no less than 100mm. Even now we are sure some of the rooms were nearer 200mm but that was what he wanted. It was necessary to blind the hardcore with sand so that the polythene sheeting laid under the concrete would not be pierced by stones.

The laying of the concrete may have only taken five evenings but it took a tremendous effort, not least from Clive's parents without whom it could not have been done. Ductings for the central heating pipes had to be made in the concrete as it was being laid, and this made it impossible to get a straight run at the concreting. Still it will have saved us work later—though it does seem odd designing a central system for a house that does not yet exist.

Whilst the dampcourse was laid Clive put the finishing touches to the benching inside the manholes. Before work could start on the house the ground around the house had to be prepared so that it could take the scaffolding. This meant the manholes had to be built up to ground level and backfilled.

On 31st August Tony the bricklayer started the blockwork. Because of the type of cavity insulation we have, the blockwork had to be built up first, the insulation placed onto the wall ties and the facing bricks then brought up. Next the patio door frames for the dining room and lounge were put in place. When the first pack was opened we noticed that a lot of the faces had white lumps in them which spoilt the look of the face. There was something wrong with them, and we phoned D & M who arranged for a rep from the brickworks to come and look at them. He told us that he would have some of them changed, and that we should sort out the bad bricks as we went along. Then he went away. The pile of bad bricks kept building up, and replacement bricks did not appear. D & M got him to come back again and we were promised a complete new load of bricks. There were delays in getting these, and in spite of all that could be done the new load did not arrive until 26th September.

Because of this delay Tony had concentrated on the internal blockwork and we could now see what the rooms were going to look like. By 18th September we were able to put the lounge joists up and all of them were in position at the beginning of the next week. Getting the heavy lintels and RSJ into place was impossible for us on our own so we enlisted the help of Mick and his forklift tractor. This got them up to the right level and we only had to put them in the right place. Not people to miss an opportunity we also got Mick to move the packs of Hemelite blocks from around the back nearer the house.

Su and Clive Mortley.

On 23rd September D & M delivered our roof trusses, barge boards, soffit, fascia boards and roof ventilation kit ready for when we needed that stage. By this time all the first floor joists had been passed by the Building Inspector and the building society architect.

We decided to try our hand at putting the roof trusses on the garage whilst the main house was still being built up and 1st October was the chosen date for this. Su's dad was invited over to help because he knew what to do and Su volunteered to take her mum shopping. The weather went down in our diary that day as being absolutely terrible with fierce wind and lashing rain and our thoughts were "thank god we did not put the main house trusses up today". We had to laugh at this later because as it turned out the weather was even worse when we did put them up.

10th October saw the delivery of a new load of bricks which now meant that our complete delivery of bricks had been replaced. It was quite a relief. The following day the roof plate was put on and the countdown began for putting the trusses up at the main house. This had to be well organised—how, we

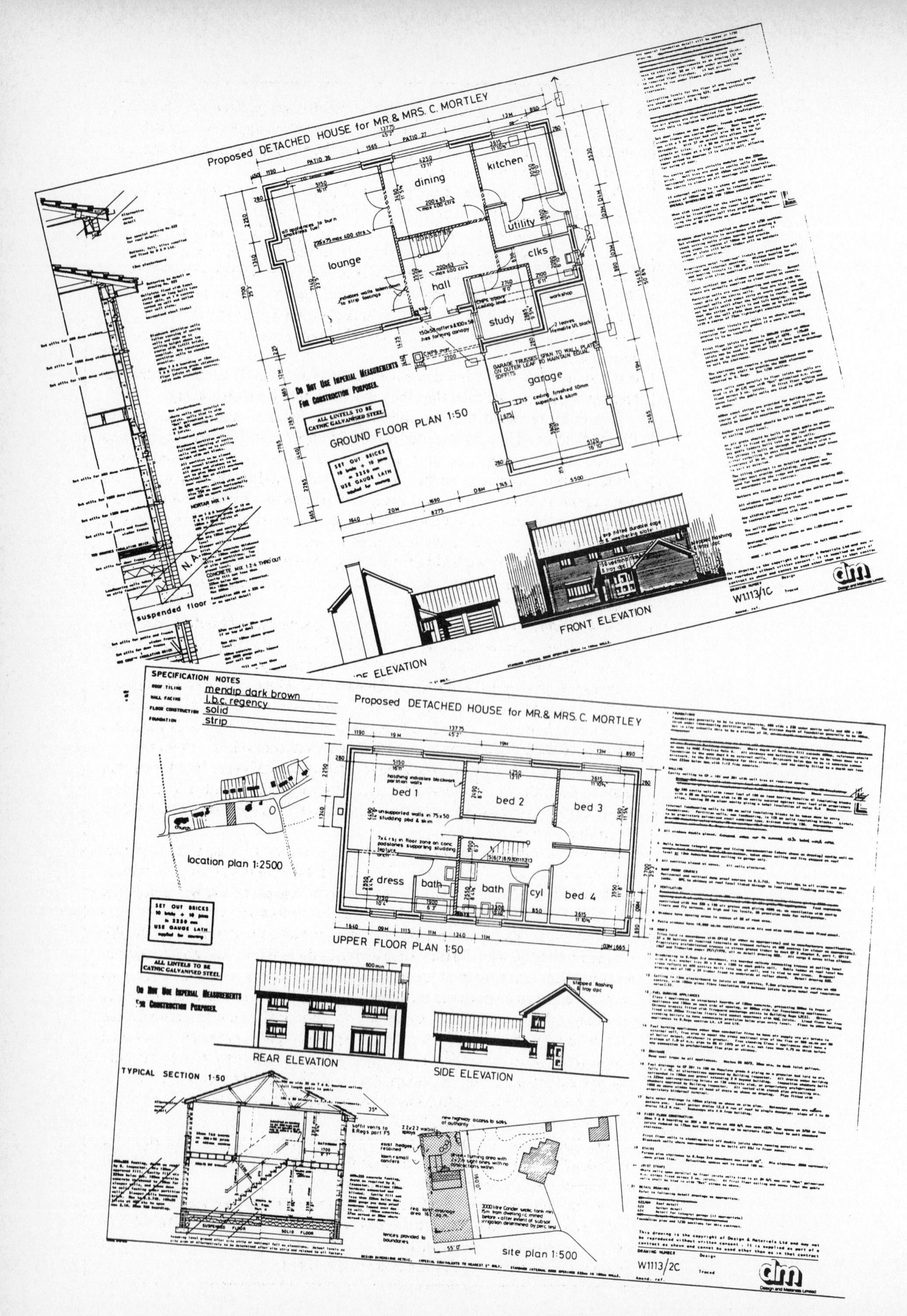

Proposed DETACHED HOUSE for MR.& MRS. C. MORTLEY
kitchen
dining
utility
lounge
clks
hall
study
workshop
garage
GROUND FLOOR PLAN 1:50
ALL LINTELS TO BE CATNIC GALVANISED STEEL
DO NOT USE IMPERIAL MEASUREMENTS FOR CONSTRUCTION PURPOSES.
SET OUT BRICKS 10 bricks + 10 joints = 2250 mm USE GAUGE LATH
FRONT ELEVATION
SIDE ELEVATION
suspended floor
MORTAR MIX 1:4
CONCRETE MIX 1:2:4
W1113/1C
dm
SPECIFICATION NOTES
ROOF TILING mendip dark brown
WALL FACING l.b.c. regency
FLOOR CONSTRUCTION solid
FOUNDATION strip
Proposed DETACHED HOUSE for MR.& MRS. C. MORTLEY
bed 1
bed 2
bed 3
dress
bath
bath
cyl
bed 4
UPPER FLOOR PLAN 1:50
location plan 1:2500
church
SET OUT BRICKS 10 bricks + 10 joints = 2250 mm USE GAUGE LATH
ALL LINTELS TO BE CATNIC GALVANISED STEEL
DO NOT USE IMPERIAL MEASUREMENTS FOR CONSTRUCTION PURPOSES.
REAR ELEVATION
SIDE ELEVATION
TYPICAL SECTION 1:50
site plan 1:500
W1113/2C
dm
Design and Materials Limited

wondered, were we going to get the roof trusses up there? They were approximately 27 foot long and the most awkward of shapes at that. What we needed was a team of workers so, with the bribery of food a suitable contingent of relatives were duly booked for 22nd October.

After the terrible weather experienced putting the garage trusses up we thought it would be impossible for the weather to be any worse when we did the main house. How wrong we were. The wind could only be described as gale force and the rain was almost impossible to see through. None of us who worked on that day will ever forget how cold, wet, miserable and tired we were. Being so high up with no protection the rain hurt our faces and managed to penetrate every gap in our clothing, soaking everything. By the evening of that day all of the trusses were up but swaying dangerously. It was difficult to anchor them in place but at the same time not make them so rigid that the wind could snap them. The wind made them appear to be made out of matchsticks and we were very worried that they would simply be blown away.

Our diary revealed that the following day, Sunday, was the last day of British Summer Time. We needed no convincing of that. The new day showed our roof trusses still to be up but leaning precariously to one side. It was pointless trying to straighten them as we knew the wind would only whip them back again. We had to wait for the weather to improve before any more could be done. For all that week the wind continued. Tony felt it would be unwise to build the gable ends up so we had to sit it out.

By 5th November we had managed to get the gable ends built up and we could put on the gable end trusses. By now the chimney was also beginning to appear above the house.

Now we had to get the roof ready for tiling, which meant putting up barge boards, barge feet, soffits and fascia boards, lead flashing and porch roof trusses. This particular stage seemed to take eternity.

Included in our package from Design and Materials was the felting, battening and tiling of the roof, and we booked the contractors to start on 7th November. This meant we had to work like lunatics to meet this date and floodlights were rigged up so that we could work through the evenings.

Finally on Friday of that week the roof tiles were delivered by Marley, but were so green that at least a quarter of them were broken in unloading. We set about arranging replacements, anxious that the tilers should not be held up when they arrived.

The weather was beginning to show signs of deteriorating again and our roof seemed so vulnerable to us. Severe frosts and the possibility of early snow were predicted for the following week. Then we were told that there was a hold up, due to work on other roofs having been held up by this awful weather. We were told that it was hoped that the tilers would be with us within the month. We exploded. Firstly, why did they keep giving us dates if they had no intention of meeting them and why did they never have the courtesy to ring us and let us know. We had given notice at the beginning of October that we would be ready for tiling at the beginning of November, which was when the first date for the tilers was fixed. We knew the weather was going to get worse and worse, and it was all terribly worrying. Finally, on 21st November the tilers arrived but only just, as their vehicle was breaking down. At any rate they finished on 25th November and all we had to do was the flashing round the chimney and at the point where the garage joined the house.

Our thoughts were now turning to the jobs to be done to be able to drop the scaffolding down one level to the first floor windows. This was mostly the staining of bargeboards, soffits and gable ends.

Next the unglamorous job of staining the first floor windows was attacked so that they would be ready to take the glass when it was delivered by D & M. We were now working towards removing the scaffolding completely, as it had become an expense we were anxious to get rid of.

Our next delivery from D & M was on its way and this included the staircase, upstairs internal walls, flooring, loft lagging, garage doors, other external doors, internal window boards, etc. The first thing we did on the day of delivery was to put the garage doors on so we could store this delivery and lock it away.

The last delivery from D & M was the glass, including the patio doors, and this came down in time for us to reach our mid-way goal. We wanted to get the house to the stage by Christmas where it was totally weatherproof and could be locked up. To achieve this meant that we had to work every evening by floodlight in the bitter cold and snow. Have you ever tried rolling putty for windows when your fingers are so cold that you can no longer feel them? The pain when they do come round can be excrutiating.

We had to make a decision whether to put the patio doors in before or after the plastering was done. The weather made us decide 'before' and we were glad we did. Despite recommendations we also did not cover them when the plastering was done and they have not marked or stained at all.

Christmas held us up in more ways than one, both financially and socially. For the first time in 5 months we stopped work on the house and had a rest, even if it was only for a week. Living in the caravan was not conducive to festivities so we went away, and it was hard to come back to the cold and damp of the caravan and start again with quite the same enthusiasm.

In the holiday between Christmas and the New Year we decided on a change of scenery. First fix plumbing. Clive's brother Bob ran his own plumbing company so he was a natural choice to help us with this and he then carried on when we returned to work. A friend Pete was also engaged at this stage, being an electrician, to put in the cables, which were easier to lay prior to the upstairs floors being put down.

Throughout the whole construction of the house we had constantly in mind that we would be living there for sometime and the upstairs floors were no exception to this. We drilled and screwed the floor sheeting just in case we ever wanted to take it up again, although we knew it would have been so much quicker simply to nail it down. We also marked all the joists on the sheeting so that we would know in future where they were.

As the first electricity cables were going in we added the main burglar alarm cables to save crawling around in the loft later. By now the electrician and the plumber were beginning to ask particularly awkward questions such as where was the cooker going in the kitchen? Which side of the bathroom was the bath to go? Many evenings of room planning were undertaken by Su to try and answer these questions, although we suspected our minds would change when we came to decorate!

We had noticed by now that we needed a goal to work towards all the time and we were now aiming for the plastering. After looking at breeze blocks for so long the thought of plastered walls was enough to spur us on. However, the jobs to be done before the plasterers could be wheeled in seemed endless. The ceilings had to be put up, the upstairs partition walls, the staircase, the window boards and the false walls surrounding the soil stack which ran up through the house. All were tedious jobs and the house was cold, dark and unsociable, especially in the evenings.

We started on the ceilings on 4th February. Clive made a special trestle to stand on and special props to hold up the ceiling boards. It took two weekends to complete the ceilings and it was surprising the difference it made upstairs. Up until then the first floor had looked like a large barn ready for a dance with the roof trusses exposed and no dividing walls. During the week beginning 19th February Clive marked out the upstairs rooms ready to start the studding walls and put up the door frames. We were using the same gyproc as we used for the ceilings, but we were not sure which way round it was supposed to go for walls—did the side with gyproc written on it face in or out? We decided in—

Nearly finished.

yes, it was supposed to face out—so, we had to size the walls we had done before we discovered this moral; if unsure always ask first . . .

Our next job was the staircase which D & M had delivered already made up. We were not putting the banisters up at this stage; simply getting the staircase in so that it could be plastered around. No sweat we thought. We thought wrong! The actual straight run of stairs was easy enough, but the bottom tread was a false mini landing, so once again the expert guidance of Su's dad was required.

It had taken two weeks to complete the upstairs walls and another week to do the staircase and we were now nearing the end of March. We were finding that with both of us working full time and trying to do all the work ourselves each job was taking so much longer than we liked.

It was Su's birthday on 21st April and we decided to combine this with a house warming party. We wanted to show the house off, so for the party we wanted the walls plastered, the ceilings artexed, the floor screeded, the central heating and fireplace working and the second fix electrics in. We only had just over a month to do it all.

Su's dad was recalled to help with the window boards and he put in all the downstairs boards. Clive boxed in the soil stack whilst this was being done. The plasterers were now able to start rendering downstairs and Clive ended up finishing off the upstairs window boards with the plasterers working around him. We had trouble with the upstairs window boards because the cavity had run out resulting in the window board not being wide enough. Clive had to make up hardwood extensions to fit on the back of each window board to shunt them out further, and small jobs like these take time. The difference that rendered, let alone plastered walls made was unbelievable. Walls start to look like walls and the ceilings look lower than they used to. The sound difference is also quite remarkable. Breeze blocks seem to absorb sound, but now everything echoed.

The plastering was given just under a week to dry out and Bob the artexer was let loose on the ceilings. Fourteen ceilings needed doing (including one in the garage, but excluding the kitchen) and he therefore needed 14 decisions about the pattern of finish required—stippled, comb finish etc. With the exception of the garage all the ceilings were being fitted with coving and the artexer was going to put that up for us at the same time. The ceilings were started on 26th March and the electrician came back the same week to put in light switches, sockets, roses, etc. The central heating system and immersion heater were also wired up.

Bob, Clive's brother, who was our plumber had been coming over in the evenings and had fixed radiators and had started putting in the bathroom suites. Apart from show these were not really necessary for the party as only the downstairs toilet was going to be used. It would be the only room with a door anyway!

The artexer and electrician left on 30th March and the same day the floor screeders arrived. It only took a morning for them to screed the whole downstairs of the house, which amazed us. We were expecting it to take days.

We had asked Tony to build our fireplace and he came along on 7th April. We had spent some time deciding on what sort of fireplace to have and we were still unsure as to the exact design when he started. We wanted a brick fireplace of some description, but were agreed that we did not want it to dominate the room or be imposing in any way. Finally we decided on a hole in the wall type fireplace, combining red brick and a light coloured marble. Ultimately a wide brass frame will be put around the hole itself, but this will have to wait a while.

The fireplace did cause some problems because after Tony had finished it did not work! More smoke came into the room than went up the chimney and it was clear that something was not quite right. After much experimentation with blocking off the hole at the front we worked out that the hole was too large. So we lowered the top of the hole, which meant re-positioning the lintel and it worked fine. The only problem was that doing this had messed up our plaster because the lintel had to be chopped out.

The party came and went with only one piece of damage despite an attendance of over 120 people. The lounge floor had crumbled badly. The screeder had to be recalled and the centre panel of the lounge floor dug out and relaid. We were as much to blame for this as anyone, for when the screeding was chipped up it could be seen that it was not properly dry.

After the party we were exhausted. We had crammed so much into the last two weeks that we felt we needed a rest physically and financially. This was not to be. On March 13th, Nigel Lawson had introduced new VAT regulations on built-in furniture in new houses and we had already bought the kitchen units. Our kitchen measures 16 ft by 12 ft and most of the walls had units of some description on them. The flat packs for the units filled the dining room and it was no mean feat putting them all together. All of the base units were fixed in first, and this enabled Clive's brother Bob to finish off the plumbing in the kitchen. This was important because he was moving to Wales on the 11th May and it would then be a long way to travel! To do this meant that we had to buy a kitchen sink (double with waste disposal section and surprisingly expensive).

The kitchen ceiling was the only one not artexed as we had decided to board it with tongue and grooved pine to be painted white. It seemed sensible to do this before the top wall units were put up so we dug into our pockets yet again and bought the wood and paint to do this. The drains were passed on 31st August. We moved into the house the next day.

On September 15th we went to Spain for a week and reflected on the work of the last year. A break was essential to re-charge our run down batteries and we knew that we were probably only three quarters through the whole project. On our return we sold the caravan for £550 and that episode of our life rolled into the sunset.

We still have a tremendous amount of work to do, but we know that this next stage will be slower. This is necessary, really, because a year of working at the pace we had puts an awful strain on you and it would be impossible to continue under that sort of pressure.

Below we show the financial picture which whilst satisfactory can never sum up the sense of pride and achievement we both feel. Even now we occasionally step back and look at the house and find it hard to believe that we actually built it; especially when we remember the ignorant couple that made that decision so long ago— lets build our own house. Now we have.

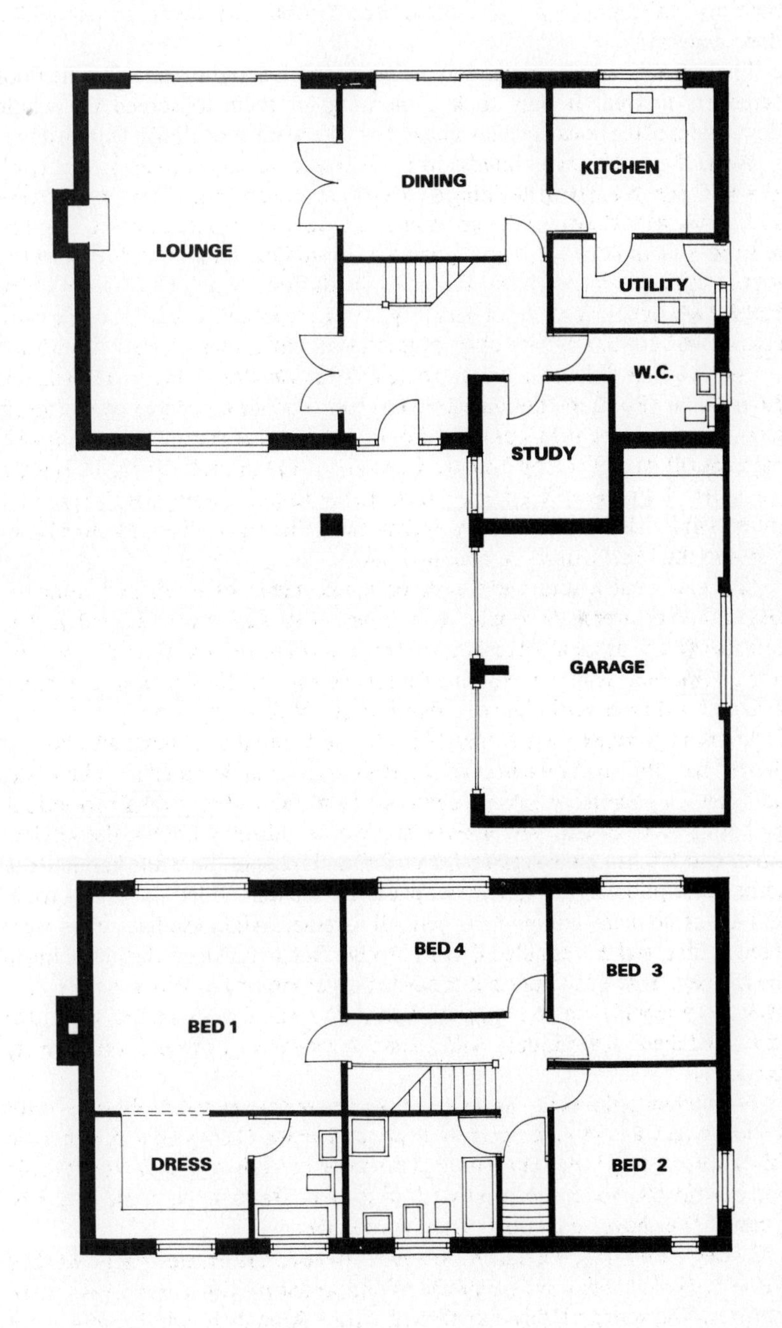

Costings

Item	£	£
Cost of moving from previous house		1030.00
Caravan plus redecoration and moving costs	1172.00	
Planning permission fee for caravan	44.00	1216.00
Cost of land 55″ × 165″		19500.00
Building society valuation fee for mortgage	71.00	
Building regulation and planning application fees	74.00	
Design and Materials for plans	109.00	
Bank arrangement fee for bridging finance	80.00	
Interest on bridging finance in total	467.00	
Building inspector fees	80.00	
Architect certificates 4 @ £69.00	276.00	
Insurances—personal and building	404.00	
Services — Electricity £170.00; Water £216.00; Telephone £78.00	465.00	
Shed	254.00	2280.00
Groundwork		
Septic tank	568.00	
JCB and lorries	570.00	
Land drains — shingle £90.00; perforated drains £100.00	290.00	
Ready mix for footings	360.00	
Foundation — bricks £610.00; labour £695.00	1305.00	
Drainage materials — cement, sand, lintels, mixer hire, bricks for manholes etc.	814.00	
Reject ballast (hardcore)	324.00	
Polythene sheet and damp course	50.00	
Hiring of vibrating plate	14.00	
Ballast and cement for oversite	409.00	4704.00
Payments to Design & Materials		
1st stage	12726.88	
2nd stage	8113.35	20840.23
Outside Costs		
Coloured sand	378.00	
Bags of cement	181.00	
Mixer hire	114.00	
Scaffolding hire	320.00	
Brickwork labour	2192.00	
Lead for roof	106.00	
Glass for front door	30.00	
Accessories for front door	30.00	
Woodstain for windows, screws and cups etc.	173.00	3524.00
Inside Costs		
Plumbing 1st and 2nd fixes	3370.00	
Electrics 1st and 2nd fixes	924.00	
Burglar Alarm	140.00	
Plasterboard for ceilings and internal walls	583.00	
Plastering and floor screeding	1200.00	
Artex and coving	699.00	
Tongue and groove for kitchen ceiling	99.00	
Fireplace — marble £81.00; bricks £50.00; Labour £60.00	191.00	
Kitchen — units and worktops £899.00; tiles £125.00; lights £14; sink £100.00; waste disposal £150.00; washing machine, tumble dryer and dishwasher, electric hob and cooker hood 955.00	2303.00	
Paints and wallpaper	60.00	
Bathrooms — two suites plus downstairs cloakroom	809.00	10378.00
Total		63472.23

From the building society valuation, necessary for the mortage, we knew that excluding decoration the house is valued at no less than £85,000. Assuming the house is decorated to the 'expected standard' the valuation is put in the region of £90–£100,000.

The above price makes no allowance for VAT, which is still to be reclaimed, but there is still some expenditure on the house to consider, including the driveway and normal decoration expenses. The amount received for the caravan has also been ignored for this same reason.

Total square footage of house and garage 2447 sq. ft.
Cost per square foot, (excluding cost of land) £17.96p

What a super story! It must be emphasized that Clive and Su are exceptional in doing so much of the work themselves. Most self-builders would have spent another £5000 to use sub-contractors for many of the jobs which the Mortleys did with their own hands — but look at the difference between the cost and the valuation!

A Timber Frame House in a Scottish Village

Duncan and Margaret Collin are teachers who have built a timber frame house in West Lothian in textbook fashion, using the best professional advice, planning everything with care, keeping to programme, and in the end achieving a 20% saving on the market value of their new home. Their story has some interesting regional slants—low land values, timber frame construction as a matter of course, and full architect's supervision throughout.

In 1983 the Collin family were ready to move to a larger home, and Duncan, having taken up teaching after a change of career from engineering, was looking for a challenge. They had a friend who had built his own house, they had read *Building Your Own Home*, and they decided to move several rungs up the housing ladder in one jump by building for themselves. The opportunity to do so presented itself when the playground of an old school was advertised for sale only a few hundred yards from their existing house. It had been stripped to the tarmac, had all services, and was priced at £8,000 for one fifth of an acre. It is on a quiet road, in a most attractive setting with views over open country; a similar plot in other parts of Britain could cost up to £40,000. They bought it and took possession in November 1983.

Meanwhile they had sent for about 30 books of plans for timber frame homes. Timber frame construction is very popular in Scotland and after carefully comparing all the options they decided on a design from Torwood Homes for a four bedroomed house of 2014 square feet with an integral single garage. Torwood quoted £14,800 for the timber frame and all the other materials usually offered by a timber frame company, plus a further £700 to erect the shell of the building. This figure included making the planning and building warrant applications.

Having got all this arranged the next step was to find an architect to issue the architect's progress certificates required by the bank and building society, and here Duncan and Margaret were lucky to find someone to deal with the certification for only £250. This arrangement developed into full architect's supervision with no increase in the fee, and they emphasise how much of their success was due to the help and advice that they got in this way. This was unusual; full supervision normally costs a four figure sum, and in general architects are wary of self builders and of involvement with standard designs from package companies. The Collin's architect was an exception, and they were very lucky. As Margaret Collin put it 'he did all the worrying for us'.

Work started in January 31th 1984, when a local site engineer set the foundations for a fee of £15. The foundations were dug with a JCB, and the concrete was poured by the family working with friends who came along to see what they were up to. The facing brickwork was handled by a bricklaying gang recommended by the architect. The original price suggested for this was £45 per 1,000 for laying common bricks and £85 per 1,000 for facing bricks. The final arrangement negotiated was £65 per 1,000 for all the brickwork of any sort. This figure will surprise English self-builders! All scaffolding, the mixer, barrows and everything else was hired separately, and the bricklayers arrangement was a true 'labour only' price.

In February the slab was cast and the bricklayers bedded the sole plate for the frame, which Torwood came and erected very quickly. The bricklayers then carried on with the external skin, and a labour-only tiler fixed the roof tiles which had been bought through a helpful local Marley representative.

Meanwhile inside the house Duncan built the internal partition walls using the studding supplied, laid the wooden floors, and installed the insulation. With the help of Margaret and the boys he took the whole job through to the point where the first fix tradesmen were required. These were all local people; the electrician was a neighbour, the plumber was the village plumber, and the heating engineer who installed the gas central heating also worked locally. When they had finished their wiring and pipe work the family fixed the plasterboarding themselves, but brought in an expert to tape the joints, skim the walls and apply an artex surface to the ceiling.

By this time everything was really taking shape and their friends began to believe that they really would make it happen. Meanwhile they found a buyer for their existing house, and were now locked in to a fixed time-table. To help speed things up a glazier was employed to install the double glazing units which had come with the shell; this was something of an extravagance but was cost effective in the circumstances. As the glazier was also a carpenter they got him to help with the stairs at the same time.

Late summer saw them completing the fitting out, installing a feature fireplace in the lounge, a wood burning stove in the living end of the huge

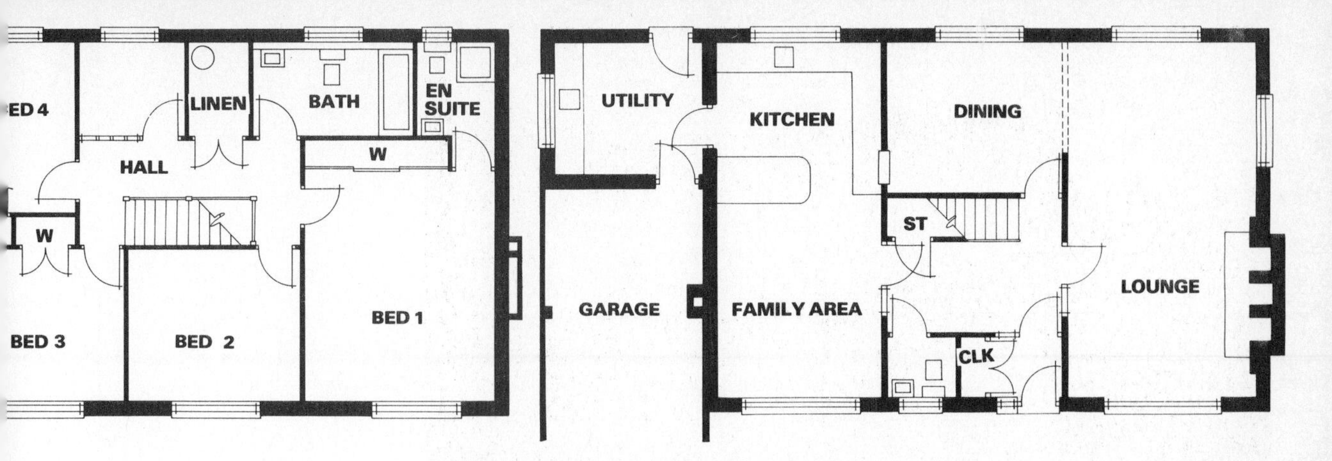

kitchen, assembling Margaret's kitchen of her dreams and finishing off two super bathrooms with acres of wall tiling. In early October they had to give possession of their old house, and so had to make a special application to the local authority for a temporary habitation certificate to enable them to move into the new home before it was completely finished. This is a formality that has long been forgotten south of the border, but the Scottish Building Warrant procedures are quite different from English and Welsh building regulation approvals.

The weekend after they had moved in the family went away for a long weekend to recharge their batteries, and to do absolutely nothing for a few days after all the pressures of the summer, when they had worked to 9.30 every evening. Like most people who choose to build on their own in their own community they are full of praise for all their friends who have helped—and insist that I give a mention to 'David and Margaret across the road who were wonderful'. I hope this acknowledgement in 30,000 copies of *Building Your Own Home* will express adequate thanks!

Margaret Collin laying flagstones at their front door.

The Devan's House in Wiltshire

Jim and Anne Devan built a house in a Wiltshire village in 1984, using *Building Your Own Home* as a text book. When they finished they wrote to the publishers offering to let their new home be featured in the 1985 edition—and here it is. This self-build project has been very much a text-book job all the way through, taking the average time (8 months), involving costs under £20 a sq. ft., and with total costs far below market value.

In 1982 the Devans, with three boys of six, three and one, decided that they wanted to move and that if possible they would build for themselves. Some years previously they had extended their first house, they had friends who had built for themselves, and they felt sure that this was something they could handle. The only snag was that they wanted to stay in the village where they were living, and that building plots were both scarce and expensive.

To meet this situation they did what is often advised and rarely done; they set out to buy a plot, instead of simply waiting for one to be offered to them. They identified and listed all the pieces of land in the village where it was conceivable that planning consent could be obtained. Then they traced the owners and asked if they would sell. At first all said "No", although none seemed irritated by the question. Then, in September 1982, they were told by the owner of a large infill site that he had actually got detailed planning consent for three houses on it, and intended to sell it at some time in the future. From this it was an easy step to actually buying the plot in January 1983.

The spring and summer of 1983 were taken up with all the preliminaries that are necessary before work can start. The existing house was sold and the family moved into a caravan on the new site. With a young family it was obviously necessary to connect the caravan to the drains and to provide a mains water supply. A mortgage had to be arranged with a building society,

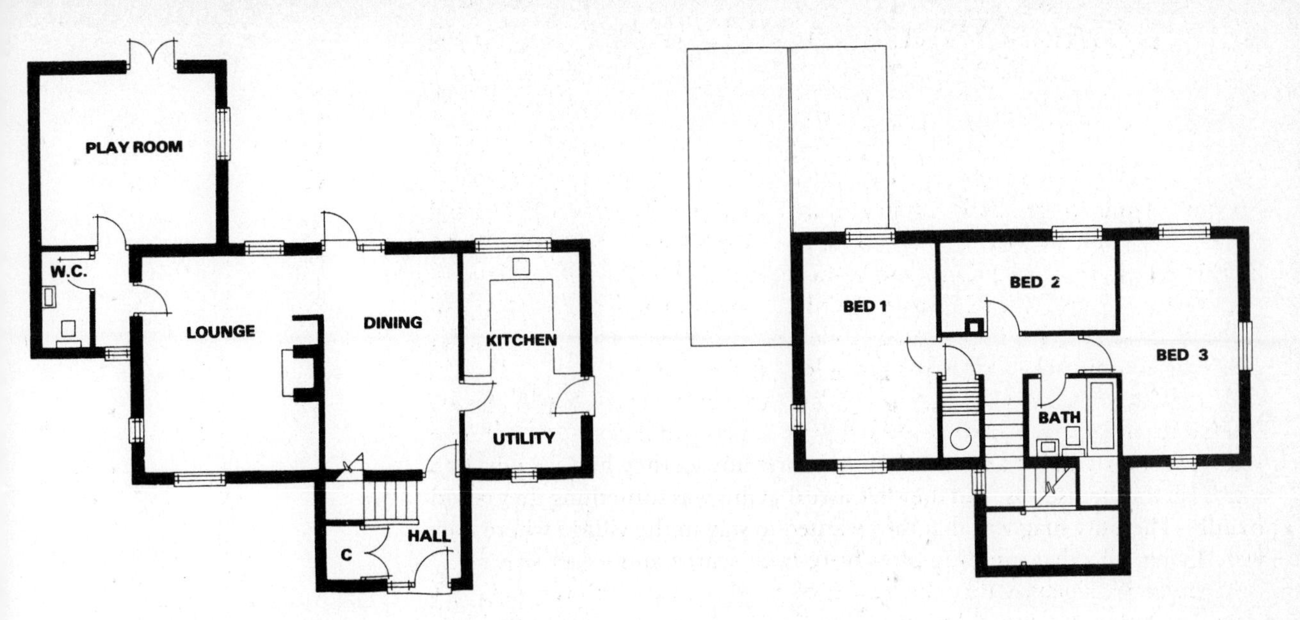

Architects Lisney Associates, Bath.

and bridging finance obtained from the bank. Finally, arrangements had to be made to use the plans which had already been approved. All this took until August 4th, when the JCB moved on to the site and Anne and Jim could start pouring their foundations and building the sleeper walls to dpc.

It often happens that one good tradesman can recommend another. The JCB driver recommended a 'two and one' bricklaying gang to take the job on from the slab on a very unusual basis. This was that the professionals would build the two skin external walls to the main part of the building, while at the same time Jim would build the internal cross walls and Anne would build the single storey playroom at the rear. This sounds like an organisational nightmare, but it all worked well, and Anne's masonry is indistinguishable from that of the bricklayers. Jim Devan then fixed his roof trusses, and tiled the roof himself. Not having a head for heights Anne concentrated on a feature fireplace in the lounge which is shown in the photograph opposite.

By now it was mid-winter, and fate decreed that there was an exceptional gale the night that the roofing felt had been fixed and before the tiles were laid. The family were living on the site, and when the felt started working loose Jim got up with it with extra battens to make it all secure at the height of the storm—not to be recommended, but typical of the sort of things self-builders do. Someone up there looked after him and he was not blown off the roof as he deserved.*

With help from friends the family handled their own plumbing, heating and electrical installations and then engaged a plastering gang at a fixed price for all the plastering work. A glazier was brought in to glaze the windows, but he was the last tradesman employed; everything else was their own work. They moved in at the end of May, 1984.

Oddly enough the part of the job which Jim enjoyed least was the scaffolding; a putlock scaffold to suit someone else's putlock holes was not easy. In the circumstances a proprietary tower scaffold might have been better. Anne's only memories that are less than happy come from living in a small caravan with three small boys on a building site in winter—the rest was all fun. Both would do it all again, and when they find the next site they probably will.

* The same storm is described in Su Mortley's diary on page 152.

Anne Devan's Fireplace

Actual Costs for House (Without Garage)	
Land	17000
Insurances	110
Fees — Solicitors	424
Fees — Inspections (Council, Building Society & Architect)	322
Site clearance & Foundation digging	505
Foundation concrete	345
Foundation blocks	305
Hardcore	189
Oversite concrete	233
Outside blockwork (incl. lintels, quoins)	1885
Inside blockwork (incl. lintels)	1440
Cavity insulation	263
Floor joists	318
Roof trusses	639
Roof tiles	493
Sand	379
Cement	360
Fitted kitchen (incl. hob & cooker)	1138
Central heating	864
All other materials & parts	6842
Cement mixer & wheelbarrow	120
Scaffold hire	362
Hire of pumps, compactor, etc.	81
Glazier	171
Bricklayer	2200
Plasterer	1260
Service connection fees	285
Total	£38533

Castle Self-Build Association in Kent

The Castle Self-Build Housing Association is one of two associations featured in this edition of *Building Your Own Home* which are managed by consultants Wadsworth and Cudd of Hurstpierpoint in Sussex. The other is the Sylvan Hill Association, which was formed by the consultants to work on a scheme for which they already had a blue print. Castle is different; it is an association that recruited Wadsworth and Cudd as their managers.

The Association was formed in 1978 with a full quota of enthusiastic members and no land. They spent three years with abortive schemes until in 1982 they narrowly missed a site in the Walderslade Development Area near Rochester in Kent. The local authority suggested that they might wish to consider another site at Walderslade, and at this point the Association members decided they wished to retain a consultant and made careful enquiries about the various possibilities. The upshot was that they retained Wadsworth and Cudd to handle their affairs at Christmas 1982, and on October 14th 1983 they started work on site.

Their development of fifteen houses is called Bellgrove Court, and is one of a series of small developments of houses of all sizes being built at Walderslade Woods. The neighbouring builders' developments give the members an opportunity to monitor local values, and, as often happens, the self-builders' homes are among the largest and most attractive in the area. They are a mix of three and four bedroomed detached houses, all of them slightly different in appearance but with the same basic layout and fittings. The current market values are around £55,000. The first one was occupied after seven months, and the whole scheme was scheduled to finish at the end of January, 1985, when the members will have been fifteen months on site.

The Association has a committee of seven which has been unchanged since they started. The wife of one of the members is the treasurer, another handles the clerical work, and all the ladies do the cleaning up on site and decorate their own houses, which is more use than most associations make of this source of help. They have other interesting features as well. Although their

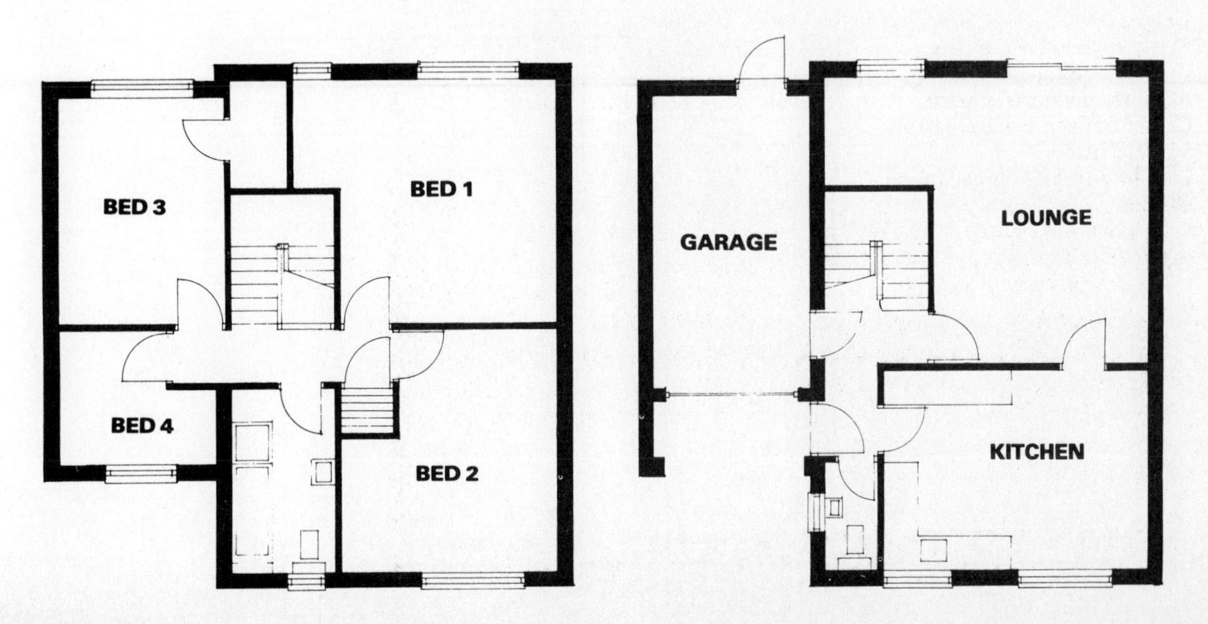

Castle Association — work in progress in July 1984

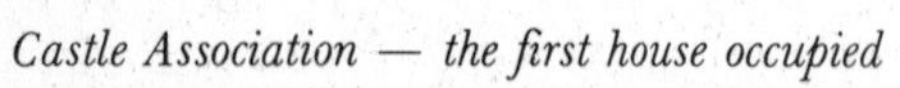

Castle Association — the first house occupied

rules make provision for fines for those who are absent, late, or simply unlucky, the Castle Association has never invoked them and feels that they can rely on members to work extra time to make up any hours lost. Many other associations would criticize this, claiming that sticking to the rules makes for easier relationships, but Castle certainly seems to be a happy association and no-one has left the group since they started work.

Castle also did another unusual thing, which was to mix all the concrete by hand for trench fill foundations. The forty tons for the foundation of each house involved all the members working together from 7.30 a.m. to 6 p.m., with two mixers and eight barrows. Some financial saving was made compared with using truck mix concrete, and elected foreman Ray Marriott also thinks it was a splendid way to settle everyone down to work together. At any rate, from these foundations the members went on to handle all the trades themselves, except for the plastering, and the first house was occupied on May 21st 1984. Members worked twenty nine hours a week, plus three full weeks a year. The normal hours are Saturday and Sunday from dawn to dusk plus Tuesday, Wednesday and Thursday evenings. Monday and Friday evenings are 'off', and this is considered important to stop members getting stale.

The cost breakdown is shown below, and it seems likely that the houses will be finished at a cost of about fifty per cent of their open market value. Finance was provided by the Sussex Building Society, with members loan capital set at £500 each plus a £5 weekly subscription.

Costings

	8 Type A Det.	*2 Type A1 Det.*	*3 Type B Det.*	*2 Type B S/D*
Plot cost, Roads, Paths, Main drainage, Main services, Legal costs, Stamp Duty, Fencing	9937.86	8343.69	7694.50	5769.24
Plant purchase, Services, connection charges	513.33	513.33	513.33	513.33
Foundations	1953.15	1940.25	1837.54	1825.44
Superstructure	3699.10	2406.48	2711.36	2519.76
Roofing	1317.95	1164.04	1111.29	1146.29
First and second Fix	2928.36	2363.46	2498.61	2498.61
Final Fix	1614.76	2173.32	1517.96	1571.96
Consultants' fees, (Managers, Architects, etc.), Loan interest, Local Authority supervision fee (Roads), Planning and Building Regulation fees	6700.00	6250.00	6100.00	5800.00
Total Development Cost	£28664.51	£25154.57	£23984.59	£21644.63
Cost saving	£20335.49	£17845.43	£17015.41	£15355.37
Market value (1983)	£49000.00	£43000.00	£41000.00	£37000.00
Percentage saving (1983)	41.511%	41.511%	41.511%	41.511%
(Now much higher)				

Appendix I
VAT Notice 719

Refund of VAT to 'do-it-yourself' housebuilders

Introduction

1. If you have built a complete new dwelling on a 'do-it-yourself' basis you may be eligible to claim a refund of VAT paid on the purchase of materials. This Notice explains the details of the scheme and tells you how to make a claim.

Advice on any aspect of this scheme may be obtained from your local VAT office, the address of which can be found in the telephone directory.

Who is entitled to claim?

2. (a) *General.* You may claim provided that you have built a complete new dwelling and that the dwelling was not built in the course or furtherance of your business. The scheme applies both to individuals and to persons who have built their dwellings in informal collaboration with others who together have made bulk purchases of some or all of the materials used. If you belong to such a group you should ask the local VAT office for advice before claiming.

(b) *Traders registered for VAT.* If you are registered for VAT and build a dwelling other than in the course or furtherance of your business you should claim refund under this scheme; you must not seek to obtain tax relief by the deduction of input tax on your normal tax returns. If you are in any doubt about the proper method of obtaining relief from tax you should consult your local VAT office.

(c) *Members of Housing Associations.* This scheme does *not* apply to persons who are members of a Housing Association formed on a self-help basis to build and sell houses to its members, which is registered as an Industrial and Provident Society, and which complies with the model rules issued by the National Federation of Housing Societies. These Associations are regarded as being in the business of building houses for sale and are liable to be registered for VAT. If registered they may recover input tax on their purchases and zero-rate their sales of new houses to members.

What kind of building qualifies for relief?

3. The scheme applies to any complete new dwelling provided it did not result from the conversion, reconstruction, alteration or enlargement of *any existing building,* (irrespective of whether that building had been used previously as a dwelling).

The following also qualify for refund, *provided they are built at the same time as the dwelling:*

(a) any built-in garage;

(b) a detached garage, provided it is occupied at the same time as the dwelling;

(c) a separate fuel store, on the same site as the dwelling;

(d) works such as paths and drives, boundary walls and fences.

Other ancillary buildings such as greenhouses or garden sheds do NOT qualify.

Must the dwelling have been built single handed?

4. No. You may have employed specialist help such as a bricklayer, plasterer, plumber or electrician. However, where such a contractor is registered for VAT purposes his services and any materials he supplied in connection with them would have been zero-rated, so there would have been no tax charged and thus none to be refunded. If he is not registered he should not have charged tax, so again no refund can be made.

Must the dwelling have been built for your own occupation?

5. No. Provided that you have not built the dwelling in the course or furtherance of any business and that all the other conditions set out in this Notice are fulfilled, you may claim a refund.

What goods are covered by the scheme?

6. To come within the scope of the scheme the goods must have been:

(a) purchased by you from a supplier registered for VAT or imported by you on payment of VAT; *and*

(b) purchased or imported by you on or after 13 November 1974 (it is the date of purchase or importation not the date of use which is important); *and*

(c) materials used in building the dwelling or in ancillary work on its garage, fuel bunker, drive etc, or fixtures of a kind normally installed by builders in new dwellings (ie articles which a purchaser of a new dwelling would normally expect to find incorporated in it).

Goods covered by the scheme include, for example: basic building materials such as bricks, breeze blocks, timber, tiles, slates, cement, sand, aggregate, plasterboard and plaster; builders' hardware such as door furniture (locks, letterboxes, etc), rainwater goods, thermal insulation and soundproofing; builders' joinery items such as door and window frames, glass and double glazing units; built-in kitchen units and split-level cookers; materials for installing water, gas, electricity and drainage services such as pipes, electric cables, sockets and light holders, and gas piping; sanitary ware such as WC bowls and cisterns, baths, washbasins, bidets, kitchen sinks, taps, shower units, shower curtain rails, hot and cold water systems; tanks, piping, radiators and other components of central heating systems; door-bells; fireplaces; fixed flooring materials of wood-block flooring, ceramic floor tiles, linoleum, vinyl tiles and cushion flooring; convector gas fires.

Goods not normally installed by builders as fixtures and therefore not covered by the scheme include, for example: air conditioning; carpets, underlay and carpet tiles; clocks; cookers or cooker hoods; curtains, curtain rails and fittings; door chimes, extractor fans (except in water closets with no other form of outside ventilation); mirrors, lamp shades and ornamental light fittings, porch lamps, light bulbs and fluorescent tubes; electric fires (unless integral parts of central heating systems); storage radiators and paraffin heaters; refrigerators; roller and venetian blinds; TV aerials; washing machines; waste disposal units; water softeners; topsoil, trees, shrubs and flowers, grass seed and turfs.

The purchase or hire of tools or equipment used for the construction of the dwelling, (such as cement mixers and scaffolding) and fees for professional and supervisory services (such as architects fees) are excluded from the scheme.

How is a claim made?

7. You can get claim forms from your local VAT offices. The forms required are:

(a) Form VAT 431 on which the claim itself is made.

(b) Form VAT 432 on which you must describe the dwelling and list the quantities of all materials used (including any supplied to you before 13th November 1974 or purchased from suppliers without a charge of VAT, even though you may *not* claim for these). See paragraph 14 for 'finishing-off' claims and paragraph 15 where a house 'kit' has been used.

(c) Form VAT 433 on which you must list goods for which the suppliers' invoices or import documents show *VAT as a separate item*. Do not include on this form any goods which are not eligible for relief (paragraph 6).

(d) Form VAT 434 on which you must list eligible goods (paragraph 6) on which VAT has been charged but *not* shown as a separate item on the suppliers' invoices. Do not include on this form any goods which are not eligible for relief.

You must make sure that you provide all the information required by the various headings of the forms as set out in the printed directions and the 'Notes on Claiming' at the end of this Notice. Inaccurate or incomplete forms which have to be returned to you for correction or completion will delay consideration of your claim.

You must also make sure that you send with the claim invoices received by you from your suppliers and import documents for any goods you have imported. Copies of these documents (including photocopies) are *not* acceptable.

Send the completed forms, invoices and import documents to your local VAT office accompanied by the documents described in paragraphs 11 and 12.

May claims be made as the work proceeds?

8. No. Only one claim may be made for each dwelling. It must be lodged after construction of the dwelling is complete.

Is there a time limit for claiming?

9. Yes. The scheme is adminstered under regulations which provide that claims must be made within a period of 3 months from the date on which the construction of the dwelling is complete, but allow the Commissioners discretion to accept belated claims where there are exceptional circumstances. In practice the Commissioners will accept claims submitted a short period after the 3 month time limit but are reluctant to do so if a claim is very late without good reason. It is in your own interest, therefore, to ensure that your claim is submitted in good time, otherwise it may be refused.

If, however, you are unable to submit your claim within the statutory time limit, then, when you do send it in, you should enclose a letter outlining the reasons for the delay.

When is a dwelling complete for the purpose of the scheme?

10. Normally, the date of completion is close to the date when the building is first occupied. However, in some cases a dwelling remains unoccupied after completion and in other cases a dwelling that is not completed but is in a habitable condition is occupied while work on the dwelling and/or its site continues.

To meet the differing circumstances, you should enter on Form VAT 431 *both* the date of completion of the dwelling (ie when work was completed on the dwelling and its site) *and* if appropriate, the date of habitation (ie when the dwelling was first occupied).

Is evidence of the construction of the dwelling required?

11. Yes. A copy of the planning permission is required together with one of the following documents:

(a) a certificate or letter of completion from the local authority; or

(b) a certificate or letter of habitation from the local authority (in Scotland a temporary certificate of habitation); or

(c) a rating valuation proposal from the District Valuer; or

(d) a certificate from a building society in the following terms:

'This is to certify that the ....Society released on..the last
(date)
instalment of its loan secured on the dwelling (and the detached garage)* at ....
.... because it then regarded that dwelling (and that detached garage)* as substantially complete'.

*Omit the words in brackets if there is no detached garage.

If a detached garage has been built in addition to a dwelling, this fact must be stated either on one of the documents (a) to (d) above or in a separate letter.

If you cannot obtain any of the documents (a) to (d) above you may provide a certificate of completion signed by an *independent* architect or surveyor. If you are unable to obtain a copy of the planning permission you should send some other evidence that permission has been granted and of the terms of the permission.

What evidence of tax charged is required?

12. You must obtain from any registered person from whom you purchased goods on which you intend to claim refund of tax, an invoice bearing his VAT registration number and detailing the quantity and description of the goods. The invoice does not have to be receipted. It will simplify the job of making your claim if VAT is shown as a separate item on the invoice, but if it is not, you may still claim the tax by calculating it as explained in paragraph 13. Invoices for supplies which are zero-rated are not to be included. The invoices which you submit must be originals; copies will not do. When your claim is examined the invoices will be stamped by the VAT office for identification purposes and returned to you in due course. If you have imported goods for use in the construction of the dwelling you must submit your copy of the import entry showing the amount of tax you have paid.

How is the tax refund calculated?

13. By completing Forms VAT 433 and VAT 434 as indicated below you will be able to arrive at the amount of VAT which you may claim.

Form VAT 433 must be used for those eligible goods for which you have obtained either invoices showing VAT as a separate amount or evidence of importation. The amount of VAT shown against each listed item is to be inserted in column (5) of the form.

Form VAT 434 must be used for goods for which you have been unable to obtain invoices showing VAT as a separate amount. The value to be shown in column (5) is the price shown on the purchase document. The VAT amount should be shown in column (6). For goods which were chargeable with VAT at 8% up to and including 17th June 1979, the tax should be calculated as 2/27 (two twenty-sevenths) of the column (5) value. For goods which have been chargeable at 15% from 18th June 1979, the tax should be calculated as 3/23 (three twenty-thirds) of the column (5) value. For goods chargeable at any other rate the VAT involved can be calculated in this way:

$$\frac{\text{Column (5) Value} \times \text{rate of tax}}{\text{rate of tax} + 100}$$

If your are still not sure how to calculate the tax your local VAT office will help you.

Instead of making a separate tax calculation for every item listed on Form VAT 434 you may total the column (5) figures for all those goods listed which bear the same rate of VAT, calculate the amount of tax on the total figure. This should be entered in column (6).

To complete your claim VAT totals from Forms VAT 433 and VAT 434 should be transferred to Part B of Form VAT 431.

'Finishing-off' claims

14. A claim may be made for goods used to 'fit-out' or 'finish off' a dwelling, after you have had the structural work completed. This does not include dwellings fully contructed by a speculative builder.

In such cases, as it may prove difficult to obtain information to complete fully the schedules, only the details of items for which a claim is being made need be listed.

Prefabricated House – Kits

15. Provided a builder's specification, which lists the items contained in the house kit, is submitted as part of the claim the schedules need only be completed to show items not included therein.

Is anything further required when making a claim?

16. Provided the claim form, declaration and schedules have been correctly completed and are accompanied by all the supporting evidence Customs and Excise will normally be able to check the validity and amount of the claim without further reference to you. However, in some cases it may be necessary to ask for additional information. You may be asked to supply copies of the plans of the dwelling; a certificate from an independent quantity surveyor or architect that the goods in the claim were, or in his judgment were likely to have incorporated into the dwelling or its site; or written evidence that a garage which is included in the claim is rated as such by the local authority.

Will the claim be acknowledged?

17. Customs and Excise will send an acknowledgement of each claim received. If you do not receive an acknowledgement within 14 days of submitting your claim you should contact the office to which it was sent without delay.

How will the refund be made?

18. After your claim has been checked and if it is found satisfactory, Customs and Excise will send you a payable order which can be presented through a bank.

Is there a right of appeal?

19. Yes. If Customs and Excise refuse to grant a refund or if you do not agree with the amount paid you may ask them to reconsider your claim. If agreement cannot be reached you have the right of appeal to an independent VAT Tribunal. There are time limits for making an appeal and these, together with the full appeal procedure, are explained in a leaflet which may be obtained from your local VAT office.

NOTES ON CLAIMING

Before submitting your claim please read this Notice carefully and check that you have covered all the following points. If you fail to provide all the documents required or to complete the forms properly delay is likely in the consideration of your claim.

The scheme provides for a refund of VAT charged only on *GOODS* and *MATERIALS* incorporated in the dwelling or its site (see paragraph 2). Tax charged on the purchase or hire of tools or equipment used in the construction of the dwelling, or on haulage charges, solicitor's and other professional fees, etc, cannot be refunded. Services provided by a contractor (eg a bricklayer or electrician) in the course of the construction of the dwelling, and any goods and materials supplied by him in connection with those services are not chargeable with VAT (they are zero-rated) and therefore there is no VAT to be refunded (see paragraph 4).

CHECK LIST

(a) Have you attached to your claim the evidence of completion of the dwelling and a copy of your planning permission? (see paragraph 11).

(b) Have you completed and signed Part C of the claim, Form VAT 431, and entered your name on all the schedules?

(c) Have you completed Form VAT 432 fully? Quantities of all goods and materials used in the construction of the dwelling must be entered whether or not you are claiming VAT on them. (see paragraphs 14 and 15 for the treatment of 'finishing-off' claims or where a house 'kit' has been used).

(d) Have you shown the quantities and described the goods and materials briefly on Forms VAT 433 and VAT 434. It is not sufficient to write 'As invoiced'.

(e) Have you included all the goods and materials on which you may claim a refund, on either Forms VAT 433 and VAT 434? (see paragraph 2).

(f) Have you excluded all items which are not eligible for a refund? (see paragraphs 3 and 6).

(g) Have you attached to Forms VAT 433 and VAT 434 the *original* invoices for all the items included in the claim, in the order in which they are listed? (see paragraph 12).

(h) Are all the invoices dated and do they bear the name and VAT registration number of the supplier? (see paragraph 12).

V.A.T regulations are subject to detail changes from time to time, and readers likely to be concerned with V.A.T. claims under the provision of Notice 719 should confirm the latest position at their local V.A.T. office.

Appendix II
S.B. Association Model Rules

The regulations detailed below under 'Building Programme' have been designed to protect the interests of all members of the Group. It is obviously important that every member knows that he will not be called upon to 'subsidise' lack of effort by others.

These regulations are suggestions only and will be discussed in detail at an early meeting and are subject to ratification by the members of the association.

membership

1. Nothing in these regulations shall apply, or be deemed to apply to any Local Authority or County Council holding a share in the Association.
2. For the purpose of filling vacancies, the committee shall cause to be kept a list of the names of persons desirous of becoming members of the Association. From this list all new members shall be elected by the Committee, who need not take the first on the list, but may take other matters into consideration.
3. The amount of Ordinary Loan Stock required under the rules of the Association to be taken up by members may be provided in instalments if necessary.
4. All payments made by members under the rules of the Association shall be made to the treasurer and to no other officer or person.

building programme

1. With the exception of the building tradesmen referred to in Paragraph 2, all members will work 14 hours per week December, January and February, 16 hours per week March, October and November and 20 hours per week April to September inclusive. Breaks, when spent on site, are included in these figures. 75% of member working hours must be worked between the hours of 7.00 a.m. and 8.00 p.m. Saturdays and Sundays. Hours worked through the week between 8.30 a.m. and 5.00 p.m. will count as week-end hours.
2. Experienced bricklayers, house joiners and plasterers will receive an allowance of two hours per week, other fully experienced site tradesmen will receive an allowance on one hour per week. These allowances are at the discretion of the Managers and may only be altered by them.
3. All members will, when starting work, sign in at the exact time and sign off when they stop work. Details are to be recorded of the exact work if done at times other than at the week-end. Members must sign off when they leave the site, unless instructed to do so by the site foreman, and members are expected to sign off when they are not working. (Breaks accepted 10 mins. morning, 30 mins. lunch break, 20 mins. tea—when part of the working day and spent on site).
4. Each member's hours will be added at the end of each month, on the last Sunday in each month, and for each hour below minimum the member will be charged £2, irrespective of any previous overtime worked, or sick hours granted. In the event of certified sickness during the last week-end of the month the members will be allowed to the end of the following month to make up the hours lost during that week-end. All fines are paid to the Association and are debited to the Members account.
5. Any time lost through site accident will not be subject to a fine unless the Association decide otherwise.
6. The Time-keeper will keep a cumulative total of each member's hours. Irrespective of any fines which may have been made, by the end of the scheme each member must be above an agreed minimum. In the event of certified sickness the fine will still apply but an allowance will be added to the member's total. The effect of this is that hours not worked on site will be fined and will also have to be made up, but although sickness hours are fined they do not have to be made up. The committee would also consider if necessary granting hours in respect of wife's illness. All fines accrue to the Association.
7. The Time-keeper will keep a list, up to date, showing the number of hours members have worked, sickness hours, penalty hours etc., and which members are above and which below the average.

8. Members will work 1 week of the annual holiday as well as their ordinary times, at a time to be agreed by the Committee and for this week the hours will be 8 hours per day. Members will also work one day extra at Spring Bank Holiday and Easter, the day to be decided by the Committee.
9. Extra Holiday Incentive Scheme—In order to help members have a longer holiday, a special bonus will operate as follows. 50% of extra hours worked between March and September (inclusive) may be used to offset extra holiday hours. The time-keeper must be informed in advance as to when the member proposes to take this bonus.
10. All members will completely specialise in one aspect of the building work as decided by the Committee, and will be expected to become completely proficient in the craft.
11. Members will not do work of other trades unless specially instructed to do so by the site foreman.
12. The Association will insure its property against fire and also insure itself against claims made upon it by third parties or members in respect of accidents for which the Association may be held responsible.
13. A personal accident policy will be taken out by the Association to cover accident of the members. Any benefit from the policy belongs to the Association, but may be given to the injured member or to his dependents. In the event of the death of a member, his next of kin has the same right to occupy and purchase one of the houses. Due to the increase in insurance premiums and because of the earnings related benefits which now apply, there will be no weekly income accident insurance. However, the members are insured for major accidents, loss of life or limb.
14. The Managers shall, prior to the start of building operations, determine the order in which completed houses shall be allocated to the members. The order of building the houses may be varied by the committee. The committee may pass over the allocation of a house to a member who may be in default.
15. Members are requested to open a Building Society account with an approved Society at the beginning of the scheme and save regularly with that Society.

equipment

1. The Association will purchase or hire certain items of equipment, and these will be maintained by the Association.
2. Members will be expected to completely equip themselves for the job they are required to do.
3. In certain cases, the Association will purchase certain items of equipment and put them into the care of various members, who will be responsible for them.

design

1. The basic design of the houses is as the approved drawings indicate and no other; standardisation is vital for economy. Certain minor internal modifications are permissable if approved by the Managers, the Association's Architects and the Building Control Department of the Local Authority. Costs of labour, Association or otherwise, will be charged to the individual and an additional 10% of the extras will be charged to compensate the Association for the losses which invariably arise when modifications are introduced.
2. If members choose to do personal work on their own houses before completion, this must not delay the main building programme; penalties may be imposed by the Committee if it does so.
3. The Managers reserve the right to amend the design in the interest of improved performance or general economy.

amendment of reguiations

1. In the event of there being any conflict between these regulations and the registered rules of the Association, the latter shall prevail.
2. Ignorance of the registered rules of the Association, and of these regulations, shall not be accepted as a valid reason for non-compliance herewith.
3. Amendment for these regulations may be made at a general meeting or a special general meeting providing two weeks notice of the resolution is sent in writing to the General Managers. A two thirds majority to amend the regulations is required.

penalties

Any member who fails to observe the working conditions can be expelled from the Association upon a resolution carried by two thirds of the members, called for the specific purpose. If he should hold a licence to occupy, proceedings will be initiated to evict such a tenant.

Appendix III
Addresses for further information

NATIONAL FEDERATION OF HOUSING ASSOCIATIONS
175 Grays Inn Road, London WC1X 8UP
Assist associations with the technicalities of registration, and with all sorts of legal advice.

THE HOUSING CORPORATION
149 Tottenham Court Road, London W1P 0BN
Makes loan funds available for approved schemes run by registered associations. These are normally administered through its Regional Offices.

THE COLIN WADSWORTH GROUP OF COMPANIES
Northfield, Snelsins Road, Cleckheaton, West Yorks BD19 3UE
Self-build management consultants. Associated companies in different parts of the country. Have a first class reputation. A number of their schemes are featured in the case histories.

'THE SELF-BUILDER' MAGAZINE
Published twice yearly by:
The Colin Wadsworth Organisation from whom it may be obtained.

DESIGN & MATERIALS LTD
Carlton in Lindrick Industrial Estate, Worksop, Notts S81 9LB
Tel. Worksop 730333
Package of architectural service and supply of materials for both individuals and associations. Many of their schemes featured in this book. Send for a wallet containing their full range of designs, prices, and up to date case histories, costing £5. They will also find you a builder if you want one.

PRESTOPLAN HOMES LTD
Stanley Street, Preston, Lancs, PR1 4AT
Tel. 0772 51628
The leading (and the biggest) company offering a package service for timber frame construction. One of their homes is featured on page 101 of this book.

ROYAL INSTITUTE OF BRITISH ARCHITECTS
66 Portland Place, London W1
Send s.a.e. for booklets on the services of architects. If you have difficulty choosing an architect the RIBA will recommend one to you if you write to them.

NATIONAL HOUSE BUILDERS' COUNCIL
58 Portland Place, London W1
Send s.a.e. for a leaflet giving their full range of invaluable publications as described in the book. These are on sale to the general public as well as to builder members of the NHBC.

PLAN SALES SERVICES LTD
Lawn Road, Carlton, Worksop, Notts. S81 9LB
Sell plans for a wide range of houses and bungalows. Their 335 page catalogue, published by Prism Press, is called Home Plans for the '80s, and is in most bookshops, or available at £9.00 using an order form at the back of this book.

THE WALTER SEGAL TRUST
6 Segal Close, Brockley Park, London SE23 1PP
See the chapter on Inner City Self-Build. They welcome enquiries.

Also by Murray Armor

HOME PLANS FOR THE 80s and PLANS FOR DREAM HOMES

Britains best selling books of Home Plans, between them presenting 630 house and bungalow designs for homes of all sizes to suit all budgets, most of them in today's traditional styles. There are also features on everything that you need to know when choosing a new home, including how to evaluate design options, the choice of materials and components, heating systems, insulation, and much else — all in the author's inimitable style.

The third edition of Home Plans for the '80s has 336 pages, 230 house and bungalow plans, 100 colour photographs, 300 line drawings — a treasure trove of information and ideas for anyone thinking about a new home.

Plans for Dream Homes is a blockbuster of 480 pages with plans of 400 homes that have been built by clients of five leading designers in different styles to suit every sort of site. They vary from five and six bedroom houses to small retirement bungalows, and the plans of all of them give all the dimensions and a wealth of construction detail.

The plans in the two books are all different designs, and the two volumes complement each other.

Full sets of drawings are available for all the designs in the books at competitive prices. Whether you are wanting to choose plans for a new home of your own, or just wanting to know more about the best of today's new houses and bungalows, these are the books you must have.

Published by Prism Press and in bookshops everywhere.

Available by post using the coupon opposite, when they will be sent with the current Ryton Books list of self-build opportunities with details of plots available and self-build groups seeking members.

Ryton Books, 29 Ryton Street, Worksop, Notts.
Telephone: (0909) 76652
Please supply books as indicated. (Please tick)

Name ....

Address ....

....

Home Plans for the Eighties — £9 inc p&p ☐ Plans for Dream Homes — £11 inc p&p ☐

(Two books deduct £1)

Enclose Cheque or quote Access No. Phone order welcome BYOH

Acknowledgements

The author wishes to express his appreciation of the help and advice received from many sources in writing this edition of Building Your Own Home, and in particular to:

Dr. and Mrs R.H. Alston
Alan Atkins
Mr. and Mrs. G. Baker
Michael Beel and the Suffolk Gardens S.B. Association
Jeff Brabban
Alan and Gwyn Brooks
Jon Broome
Melvyn Butler of Sutton in Ashfield Estates Dept.
Ian and Sue Case
The Castle Housing Association
Duncan and Margaret Collin
Ted Cowling
Ian Culley
Jim and Anne Devan
Mr. and Mrs. Les Faulkener
Dr. Rod Hackney
Bruce Heath, Wadsworth and Heath Ltd
Keith and Pat Helliwell
Chris and Sue Jackson
Alan Longley
Mr. H. Mason, Chesterfield Estates Dept.
Mike Matthews and the Underwood Housing Association
Mike and Barbara Mawer
Clive and Su Mortley
The Oakwood Housing Assn
Keith and Liz Pattison
Beverley Pemberton
R.L. Photography
Len Rees
Neil and Laura St. John
Cynthia Scott
Roy and Sheila Sharp
Tim Skelton of the Milton Keynes Development Corporation
David Snell
The Sylvan Hill Housing Assn
Tom Taylor
Colin Wadsworth
Andrew Withers

A special self-builders insurance policy is available from Norwich Union, using the proposal form overleaf. Alternatively contact DMS Services for a separate proposal form if you do not want to remove this from the book.

Agency: DMS Services Ltd   Agency Reference: 2GA479   Policy No.

Proposal - BUILDING OWN PRIVATE DWELLING
The Insurer: Norwich Union Fire Insurance Society Limited

Name of Proposer

Full Postal Address

....

Postcode

Address of property to be erected

....

Postcode

Commencing date of insurance

Important - Please give a definite answer to each question (block letters) and tick appropriate boxes

		Yes	No	If "Yes" please give details
1.	Have you made any other proposal for insurance in respect of the risk proposed?	☐	☐	
2.	Has any company or underwriter declined your proposal?	☐	☐	
3.	Have you ever been convicted of (or charged but not yet tried with) arson or any offence involving dishonesty of any kind (eg fraud, theft, handling stolen goods)?	☐	☐	

		Yes	No	If "Yes" please give details
4.	Site details (a) Is the site on made-up ground?	☐	☐	
	(b) Is there any history in the area of subsidence, flooding or vandalism?	☐	☐	

		Yes	No	(If "No" please refer to DMS Services Ltd) Phone 0909 76652
5.	Will the property be			
	(a) of standard brick, stone or concrete construction, roofed with slates, tiles, asphalt, concrete or metal?	☐	☐	
	(b) occupied as your permanent residence on completion	☐	☐	

6. Will the total value of plant and equipment used exceed £2,000 on site at any one time?

Yes ☐ No ☐ If "Yes" please state:

Type and value of own plant/equipment	hiring charges incurred in respect of hired plant

7. State estimated value of building work on completion at builder price for reinstatement £

N.B. This will be the limit of indemnity for item (a) of the Contract Works Section

8. Material facts - state any other material facts here. Failure to do so could invalidate the policy. A material fact is one which is likely to influence an insurer in the assessment and acceptance of the proposal. If you are in any doubt as to whether a fact is material it should be disclosed to the insurer.

Note: 1. You should keep a record (including copies of letters) of all information supplied to the insurer for the purpose of entering into the contract.
2. A copy of this proposal form will be supplied by the insurer on request within three months of completion.

Declaration To be completed in all cases

I desire to insure with the insurer in the terms of the Policy used in this class of insurance. I warrant that the above statements and particulars are true to the best of my knowledge and belief and that I have not withheld any material information. I agree to give immediate notice to the insurer of any alteration to the circumstances described herein and that this proposal shall form the basis of the contract between us.

Date	Proposer's signature

Send completed form to D.M.S. Services Ltd., Orchard House, Blyth, Worksop, Notts. S81 8HF together with a cheque made payable to the Norwich Union Fire Insurance Society Limited.
Any queries to DMS Services, Phone 0909 76652

Norwich Union Fire Insurance Society Limited. Registered in England No. 99122. Registered Office Surrey Street, Norwich NR1 3NS
Member of the Association of British Insurers   Member of the Insurance Ombudsman Bureau

DETACH